ANTS OF NORTH AMERICA

ANTS
OF NORTH AMERICA
A Guide to the Genera

**Brian L. Fisher
and Stefan P. Cover**

Illustrated by Ginny Kirsch and Jennifer Kane
Color images created by April Nobile

UNIVERSITY OF CALIFORNIA PRESS

University of California Press, one of the most distinguished university presses in the United States, enriches lives around the world by advancing scholarship in the humanities, social sciences, and natural sciences. Its activities are supported by the UC Press Foundation and by philanthropic contributions from individuals and institutions. For more information, visit www.ucpress.edu.

University of California Press
Oakland, California

Library of Congress Cataloging-in-Publication Data

Fisher, Brian L., 1964–
    Ants of North America : a guide to the genera / Brian L. Fisher and Stefan P. Cover ; illustrated by Ginny Kirsch and Jennifer Kane; color images created by April Nobile.
        p. cm.
    Includes bibliographical references and index.
    ISBN 978-0-520-25422-0 (pbk. : alk. paper)
    1. Ants—United States—Identification. 2. Ants—Canada—Identification. 3. Ants—United States—Pictorial works. 4. Ants—Canada—Pictorial works. I. Cover, Stefan P., 1952– II. Title.

QL568.F7F48 2007
595.79′6097—dc22   2007002241

Manufactured in China
25   24   23   22   21
10   9   8   7   6   5

Cover: top, *Cephalotes varians;* row left to right, *Pseudomyrmex gracilis, Pogonomyrmex badius, Atta mexicana, Camponotus ulcerosus, Acromyrmex versicolor, Labidus coecus.* Photos by April Nobile.

No living person has exerted a more profound or more beneficent impact on the science of ant systematics than Barry Bolton. During his career, he has been a pioneer, a leader, an exemplar, a mentor, and a friend to a whole generation of ant systematists. The quality of his work sets the standard for the entire field. Therefore, it is with the greatest pleasure that we dedicate this book to him in acknowledgment of his retirement from the Natural History Museum in London, England in 2004.

———————

CONTENTS

PREFACE

Each summer since 2001, the authors have organized a 10-day, total-immersion ant research workshop called Ant Course (www.antweb.org). The course, held at the Southwest Research Station in the Chiricahua Mountains of southeastern Arizona, has made us acutely aware of the need for a simple, reliable, illustrated key to North American ant genera. For this reason, creating a workable key has become a high priority for the course.

This new generic key addresses a broader need as well. The aspiring student of North American ants arrives at a peculiar time in the fauna's taxonomic history. For many years, W. S. Creighton's (1950) pioneering and magisterial *Ants of North America* was the standard reference. Over the past 55 years, this monograph has grown increasingly outdated, a process accelerated by a resurgence of interest in North American ants during the last couple of decades. Since 1985, many important taxonomic studies have been published, and gifted collectors have demonstrated that the true size of the North American ant fauna is far larger than the 585 taxa reported by Creighton; at present, the list is nearing 1,000 species. Despite this encouraging progress, the North American ant fauna remains difficult to work with, especially for beginners. Important taxonomic resources are scattered widely in the scientific literature and vary in their approaches, descriptive language, quality, and accessibility. Even the two commonly available generic keys (Cover, in Hölldobler and Wilson 1990; Bolton 1994) are daunting to use because they lack illustrations linked to key couplets and employ too much arcane terminology.

We need a new unified monograph of the North American ant fauna that reflects modern standards in taxonomic description and analysis, applies natural history information to taxonomic judgments, and is easy to use—in short, a 21st-century Creighton's *Ants of North America*. Such a synthesis will require much more than merely reshuffling judgments, keys, and verbiage from obsolete past literature. Rather, it will require that each genus be freshly reviewed and new keys written in light of the recent collections that have expanded and transformed our understanding of the fauna.

We hope the key you now hold will be a small down payment on this larger undertaking. Our key-writing philosophy is straightforward; the function of taxonomic keys is to help identify the beast at hand. Anything else is superfluous. So we have tried to heed Henry David Thoreau's excellent advice: "Simplify, simplify!" We have minimized the use of technical language and illustrated the key couplets with clear line drawings that reduce the need for many written character descriptions. Supporting materials provide some natural history context and guidance as to the references needed to make a species-level identification.

We would like to thank the instructors and students of previous North American Ant Courses for their ruthless and insightful critiques of earlier versions of this key: they are Gary Alpert, Lloyd Davis, Mark Deyrup, Bert Hölldobler, Bob Johnson, Mike Kaspari, Jack Longino, Bill MacKay, Raymond Mendez, Hamish Robertson, Zachary A. Prusak, Roy Snelling, Andrew V. Suarez, Howard Topoff, James Trager, Walter R. Tschinkel, and Phil Ward. If this key works as we hope it will, we owe them all—our friends and fellow members of the Ant Mafia—tremendous gratitude. As for errors and omissions, they are ours alone, but at least we will be able to blame each other (one of the unsung advantages of co-authorship). We also owe a great debt of gratitude to Ant Course Advisor E. O. Wilson, who has generously supported the course since its conception. Finally, we are indebted to Kate Hoffman and associates at UC Press for their passionate response to this work. Their attention to detail and advice at all stages of production greatly improved this guide.

INTRODUCTION

There is something profoundly fascinating about ants, even when they are being a nuisance. In large part, this is because they do so many things that remind us of ourselves, and have been doing them for over 100 million years. Like humans, ants are social, living exclusively in highly organized societies that evolved originally from family groups (in the case of ants, the group consists of a mother and her offspring). Like humans, ants exhibit a seemingly endless variety of complex social behaviors. Ants were the first herders, agriculturalists, and food storage experts. Some ants fight vicious territorial wars, some "enslave" other ants, and in North America alone, over 20 species of predatory army ants march in leaderless packs on perpetual campaign. On the brighter side, they take excellent care of their mothers and their sisters, and even their brothers and sons—who do little but eat and never help with the chores.

At a time when the majority of Americans live in cities, and wildlife sightings have become increasingly rare, ants still offer a link to unbridled nature. They invade our kitchens and interrupt our picnics. They riddle our house timbers with holes, disturb our garden soils, gnaw computer cords, demolish our leftovers, and tromp through our refrigerators oblivious to the cold. These intrusions have been intensified by a small number of exotic species (such as the Argentine Ant and the Imported Fire Ant) brought to North America by humans. Some of these species have spread to infest enormous areas—further enhancing the reputation of ants as pests.

Sadly, few comprehend the vital importance of ants to the ecosystems that sustain human life on this planet. In North America, for example, close to 1,000 species of ants play essential roles in the proper functioning of nearly all terrestrial ecosystems. They are prominent agents in the breakdown of organic matter, nutrient cycling, soil turnover and aeration, seed dispersal, seed consumption, and plant protection. If ants went on strike and ceased their ecological services, the consequences would be profoundly disruptive to the natural world—and eventually tragic for humanity. The ecological importance of ants is tied to their abundance. In most terrestrial habitats, the ants on a typical acre of land outweigh any other comparable group of invertebrates. As a result, ants have the potential to help us monitor and assess the health of ecosystems. Changes in the ant community over time may signal broader changes in an ecosystem and provide clues as to probable causes. In addition, measuring the presence and abundance of invasive ants may prove to be especially valuable as an indicator of ecosystem disruption. Developing ants as monitoring "tools" is all the more urgent, given the extraordinary rate at which we are destroying natural environments and our rudimentary capacity to measure and understand the consequences of these changes.

Ants are a spectacular ecological and evolutionary success story. About 12,000 species have been named to date—perhaps half the total number of ant species occurring on the planet. The majority live in the tropics or subtropics, but though less diverse, the ant faunas of temperate areas are no less complex or interesting. The North American ant fauna is a perfect example. Ants abound in nearly all terrestrial habitats south of the Canadian tundra. The majority of our ants prefer cool to warm temperate climates and show reasonably close relationships with their Eurasian counterparts. These include most species of important genera like *Formica, Lasius, Myrmica,* and *Leptothorax.* In the desert southwest and in California, much of the fauna is essentially Mexican in origin. These include the bulk of our species of *Pogonomyrmex, Pheidole, Myrmecocystus, Neivamyrmex,* and *Crematogaster.* In addition, a small number of truly tropical genera reach Texas, Arizona, or the southeastern states (e.g., *Pachycondyla, Acanthostichus, Labidus,* and *Cephalotes*). Lastly, humans have introduced a number of exotic ants unintentionally. Nearly all are tropical or subtropical in origin and are found most commonly along the Gulf Coast or in California. Only a tiny handful are cool temperate in distribution.

A special feature of the North American ant fauna is the comparatively high incidence of social parasitism. A social parasite is an ant that must secure its survival and reproduction by integrating itself behaviorally and chemically into colonies of a related species (usually a congener). Parasitism is common in North American ants (also in Eurasian ants), but quite rare in the tropics, a phenomenon that has provoked much speculation. Just over 14 percent of our fauna consist of suspected or proven social parasites, and new ones are discovered every year. In a phylogenetic sense, social parasitism is distributed very unevenly. While most large ant genera contain a few parasitic species, 69 percent of our social parasites belong to just two formicine genera: *Formica* and *Lasius*. Social parasites exhibit an impressive and fascinating array of life-histories that are poorly documented on the whole. Dependence on the host may be temporary (e.g., occur during colony founding only) or permanent (as in inquilines and many slave-making ants). Social parasites have evolved many features apparently related to their parasitic existence, notably a general reduction in the size of the queen caste, the frequent occurrence of specialized morphology or pilosity ("hairs"), and, in some cases, the reduction or loss of the worker caste. Some social parasites are quite common. Certain *rufa* group *Formica* form enormous, polydomous (multiple nest) colonies that dominate extensive areas in the Great Plains and Rocky Mountains. *Lasius* of the *umbratus* and *claviger* species groups are abundant throughout much of North America, except in warm desert habitats. These ants have large colonies and tend aphids and coccids on plant roots, but they are seldom seen because the workers are almost exclusively subterranean.

North American ants exhibit many other fascinating phenomena. In the southern and eastern United States, fungus-growing ants collect plant material and caterpillar droppings to culture their own special fungi. Seed-harvesting ants are common in desert and grassland habitats where stored seeds form an invaluable food reserve in times of drought. Ants that build special mounds of soil and plant materials to incubate their young are widely distributed in the cooler parts of the region. Ants that tend aphids, coccids, and membracids to "milk" them for sweet fluids are found everywhere, both above and below the soil surface. We have ants that use "trap-jaw" mandibles that function much like mousetraps to capture prey; "honeypot" ants that store liquid food in the bodies of certain worker ants; solitary hunters; blind underground armies; minuscule thieves; moochers that sponge off of other ants; ants with highly specialized diets; and ants that will eat everything but the kitchen sink ... the stories are endless and there are many more to be discovered.

Entering this fascinating world begins by learning to recognize the ants you might see or collect in the field in North America. It is a sad truth about humans in general that we do not see, understand, or value what we do not perceive directly for ourselves. The forest is simply a featureless wall of green unless we begin to recognize the plants and animals which it is composed of—a learning process that reveals much of its beauty as well as its complexity. There is all the difference in the world between reading about slave-making ants or watching them on television—and realizing that the ants running across the sidewalk in front of your house are conducting a slave raid! Making this all-important transition is what this book is about. We hope to turn your focus to the earth at your feet, to the litter on the forest floor, so you can see—really *see*—the ants that share every nook and cranny of the planet with us. Each species tells a different story, each species plays a different role, and all deserve our respectful attention as our neighbors and companions here on planet Earth. You can begin the journey by learning to use the keys in this book.

KEY TO NORTH AMERICAN
ANT GENERA BASED ON
THE WORKER CASTE

1 Upper plate (tergite) of the last abdominal segment (pygidium) flattened and with a pair of distally converging rows of spines or peg-like teeth **(A)**. (Cerapachyinae) **2**

— Pygidium rounded and without a pair of distally converging rows of spines or peg-like teeth **3**

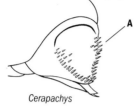

Cerapachys

2 (1) Antenna 12-segmented. Scape greatly flattened along entire length. Antennal socket without lateral carina. Frontal carinae slightly expanded laterally and sometimes covering part of the antennal insertions **(A)** . . . . ***Acanthostichus***

Note: Not commonly collected.

— Antenna 11-segmented. Scape not flattened. Antennal sockets bordered by sharp lateral carina. Frontal carinae not expanded; antennal sockets exposed **(AA)** . . . . . . ***Cerapachys***

Note: Not commonly collected.

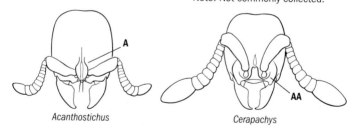

Acanthostichus   *Cerapachys*

3 (1) Postpetiole absent; waist consisting of a petiole (abdominal segment 2) **(A)**. In some cases, petiole greatly reduced and node vestigial or absent *(Tapinoma, Technomyrmex)* **(B)**. In profile, abdominal segment 3 not markedly smaller in size than abdominal segment 4 and either confluent with **(C)** or separated from it by a slight or moderate constriction **(D)** **4**

— Postpetiole present; waist consisting of two segments **(AA)**. Following the petiole is the postpetiole (abdominal segment 3), which is much smaller than abdominal segment 4 and separated from it by a strong constriction **32**

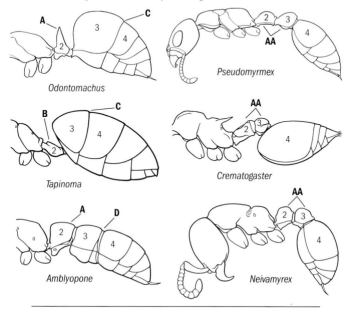

Odontomachus

Pseudomyrmex

Tapinoma

Crematogaster

Amblyopone

Neivamyrex

4 (3) Sting absent. No constriction present between abdominal segments 3 and 4 **(A)** **5**

— Sting present, often prominent. Usually with a visible constriction between abdominal segments 3 and 4 **(AA)**; if without a visible constriction (*Anochetus, Odontomachus*) **(BB)**, then mandible inserted in the middle of the front margin of the head **(CC)** **20**

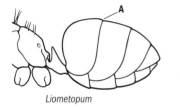

Liometopum

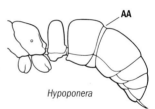

Hypoponera

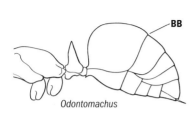

Odontomachus

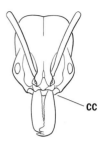

Odontomachus

5 (4) Apex of abdomen with a circular opening (acidopore), usually surrounded by a fringe of hairs **(A)**. If the fringe of hairs is not present, then the antennal insertions are set well back from the posterior border of the clypeus *(Campono-tus)* **(B)**. (Formicinae) 6

— Acidopore absent. Apex of abdomen with a transverse slit-like orifice, not surrounded by a fringe of hairs **(AA)**. Antennal insertions set at or directly adjacent to the posterior border of the clypeus **(BB)**. (Dolichoderinae) 13

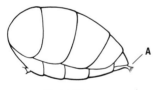

Paratrechina

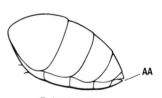

Technomyrmex

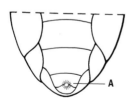

Paratrechina

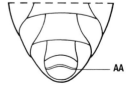

Technomyrmex

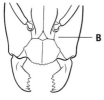

Camponotus

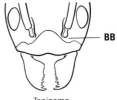

Tapinoma

6 (5) Antenna 12-segmented, eyes well developed **(A)**. 7
— Antenna 10- or 11-segmented. Eye minute, nearly vestigial **(AA)**. . . . . *Acropyga*
  Note: 1 sp. *A. epedana,* not commonly collected, Arizona.
— Antenna 9-segmented, eyes well developed **(AAA)**. . . . . . .
  . . . . *Brachymyrmex*
  Note: Not commonly collected outside southeastern
  North America.

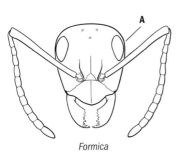

Formica

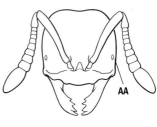

Acropyga

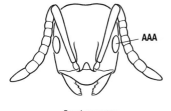

Brachymyrmex

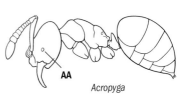

Acropyga

7 (6) Antennal sockets located well behind the posterior clypeal margin **(A)**. Acidopore lacking fringe of erect hairs. Mesosomal profile often continuous and convex. Metanotal suture rarely impressed **(B)**. . . . . ***Camponotus***

— Antennal sockets situated at or adjacent to posterior clypeal margin **(AA)**. Acidopore with circular fringe of erect hairs. In profile, propodeum distinctly depressed below level of promesonotum and metanotal suture always impressed, forming a transverse groove across mesosomal dorsum **(BB)**. **8**

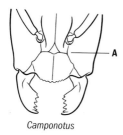

Camponotus

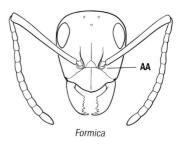

Formica

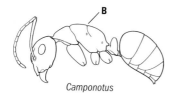

Camponotus

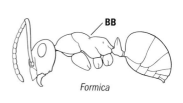

Formica

8 (7) Mandible long and sickle-shaped, with minutely serrate inner margin but lacking dentate cutting margin **(A)**. Reddish. . . . . ***Polyergus***

Note: Not commonly collected.

— Mandible not long, approximately triangular, with oblique or transverse multidentate cutting margin **(AA)**. Color variable. **9**

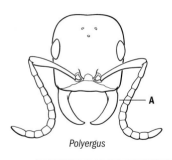

Polyergus

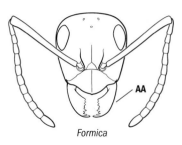

Formica

9 (8) Maxillary palpi greatly elongated, segment 4 much longer than 5 + 6 **(A)**. Psammophore frequently present **(B)**. . . . . *Myrmecocystus*

— Maxillary palpi normal to very short, segment 4 (when present) no longer than 5 + 6 **(AA)**. Psammophore absent, although short erect hairs may be present on ventral surface of head **10**

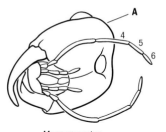

Myrmecocystus

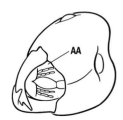

Lasius (claviger group)

Myrmecocystus

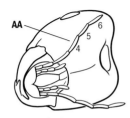

Lasius

10 (9) Mandibles with seven or more teeth or denticles. Dorsum of propodeum often longer than posterior face (declivitous face) **(A)**, or propodeum evenly rounded with faces not distinguishable. Ocelli present and well developed **(B)**. Propodeal spiracle elliptical to broadly oval . . . . ***Formica***

— Mandibles with five or six teeth or, if seven or more, then dorsum of the propodeum is notably shorter than the posterior surface (declivitous face) **(AA)**. Ocelli very small (e.g., *Paratrechina longicornis*) or absent. Propodeal spiracle circular to subcircular **11**

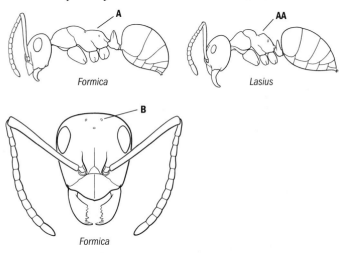

Formica   Lasius

Formica

11 (10) Antennal scape surpassing posterior lateral margin of head (if at all) by less than one third of its length **(A)**. Pronotum without black coarse hairs but with pilosity that is long or short and often golden in color **(B)**. Mandibles with seven to eight teeth or denticles. . . . . ***Lasius***

— Antennal scape usually surpassing posterior lateral margin of head by at least one third of its length, sometimes more **(AA)**. Erect hairs on dorsum of pronotum either coarse, long and black or brown, or fine and golden in color **(BB)**. Mandibles with five or six teeth **12**

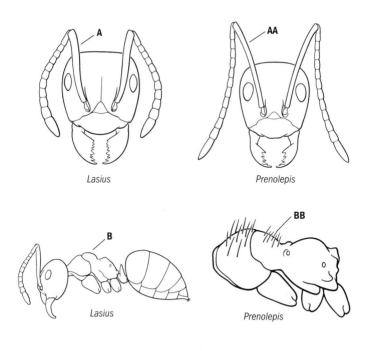

Lasius

Prenolepis

Lasius

Prenolepis

12 (11) Mesosoma, in dorsal view, strongly constricted immediately behind pronotum **(A)**. Erect hairs on mesosomal dorsum slender, golden or brownish, not coarse and bristle-like, not occurring in pairs on the mesosomal dorsum **(B)**. Scape and tibia lacking erect hairs (short pubescence present). With head in full-face view, most or all of eye posterior to middle of sides **(C)** . . . .
. . . . ***Prenolepis***

Note: 1 sp. *P. imparis,* widespread.

— Mesosoma, in dorsal view, only slightly constricted immediately behind pronotum **(AA)**. Erect hairs on mesosomal dorsum coarse, bristle-like, often dark brown or black, occurring in pairs on the mesosomal dorsum **(BB)**. Scape and tibiae often with erect hairs. With head in full-face view, most or all of eye at or anterior to middle of sides **(CC)** . . . . ***Paratrechina***

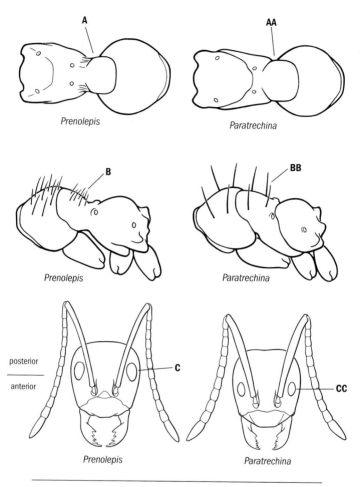

A — Prenolepis
AA — Paratrechina

B — Prenolepis
BB — Paratrechina

posterior
anterior

C — Prenolepis
CC — Paratrechina

13 (5) Propodeum (lateral view) strongly hollowed out behind, forming a shelf that overhangs the petiole, the
node of which fits snugly into this concavity. Propodeum rounded at juncture of dorsal and declivitous
face, often strongly sculptured, with large, shallow punc-
tures **(A)** . . . . ***Dolichoderus***
    Note: Eastern and midwestern North America.

— Propodeum (lateral view) usually not concave **(AA)**. If posterior face is concave in side view, then propodeum is angulate at juncture of dorsal and declivitous faces, not forming an overhanging shelf as described above *(Ochetellus glaber)* **(AAA)**. Propodeum never strongly sculptured **14**

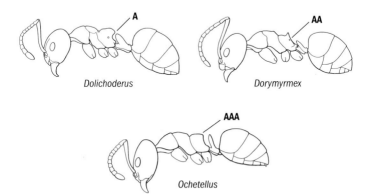

Dolichoderus

Dorymyrmex

Ochetellus

14 (13) Propodeal angle (lateral view) with a distinct dorsally projecting cone or tooth at juncture of dorsal and posterior faces **(A)**. Maxillary palp segment 3 elongated, about equal to combined lengths of segments 4 through 6 **(B)** . . . . ***Dorymyrmex***
— Propodeum without a dorsally projecting cone or tooth **(AA)**. Segment 3 of maxillary palp short, about as long as segment 4 **(BB)** **15**

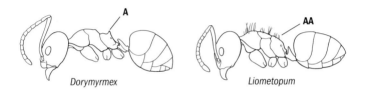

Dorymyrmex

Liometopum

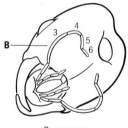

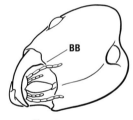

Dorymyrmex   Liometopum

15 (14) Mesosoma dorsum (lateral view) without erect hairs **(A)** **16**
— Mesosoma dorsum with erect hairs **(AA)** (most easily seen against a dark background). **18**

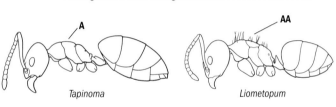

Tapinoma   Liometopum

16 (15) In side view, propodeum with a weakly concave posterior face that meets the dorsal face at a distinct angle **(A)** . . . . *Ochetellus**

Note: 1 sp. *O. glaber*, Florida.

— Propodeum rounded at juncture of dorsal and posterior faces **(AA)** **17**

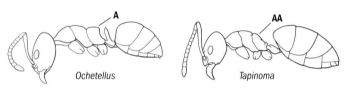

Ochetellus   Tapinoma

Note: Asterisks (*) denote introduced genera.

17 (16) Petiolar scale well developed **(A)**. First two antennal segments beyond scape equal in length **(B)**. In side view, propodeum rounded with dorsal and posterior face approximately equal in length **(C)** . . . . ***Linepithema****

Note: 1 sp. *L. humile,* widespread.

— Petiole flattened, without a conspicuous, dorsally protruding scale (petiole often concealed by the succeeding abdominal segment) **(AA)**. First antennal segment beyond scape about twice as long as second **(BB)**. In side view, dorsal face of propodeum much shorter than posterior face **(CC)** . . . . ***Tapinoma***

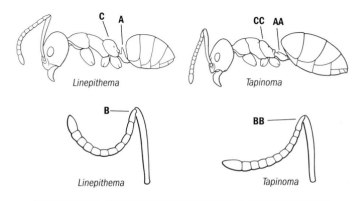

Linepithema

Tapinoma

Linepithema

Tapinoma

18 (15) Head, body, legs except tarsi dark brown-black, tarsi contrastingly pale; sides of mesosoma conspicuously microreticulate **(A)**. . . . . ***Technomyrmex****

Note: 1 sp. *T. difficilis,* Florida, greenhouses.

— If blackish, tarsi not contrastingly pale; sides of mesosoma not conspicuously sculptured **(AA)** **19**

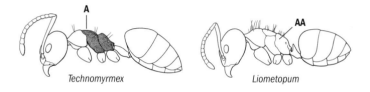

Technomyrmex

Liometopum

19 (18) Clypeal margin more or less straight **(A)**; setae of anterior margin of clypeus short and straight, ending far short of the anterior margin of closed mandibles **(A)**. Workers somewhat polymorphic. Metanotal groove reduced to thin suture across dorsum that does not clearly interrupt mesosomal profile **(B)** . . . . *Liometopum*

Note: Western North America.

— Clypeal margin convex **(AA)**; anterior clypeal margin with several conspicuously curved setae that extend to or beyond anterior margin of closed mandibles **(AA)**. Workers monomorphic. Profile of mesosoma interrupted by metanotal groove, clearly dividing mesonotum from propodeum **(BB)** . . . . *Forelius*

Note: Widespread, especially in southern North America.

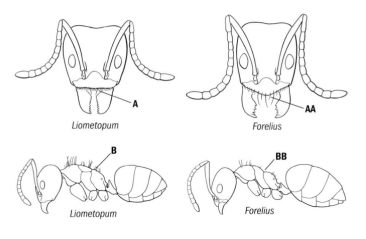

Liometopum

Forelius

Liometopum

Forelius

20 (4) Anterior margin of clypeus denticulate **(A)**. Petiole broadly attached to third abdominal segment **(B)**; petiole with distinct front and dorsal faces but without a distinct sloping posterior face; petiole separated at most by a shallow impression from third abdominal segment (Amblyoponinae). **21**

— Anterior margin of clypeus not denticulate **(AA)**. Petiole narrowly attached to third abdominal segment **(BB)**;

petiole with distinct front, top, and posterior faces; petiole's attachment to the third abdominal segment narrow and strongly constricted **22**

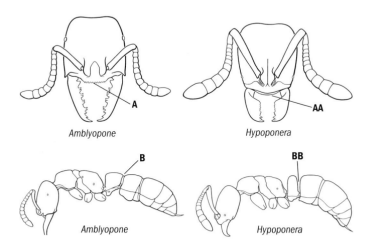

Amblyopone

Hypoponera

Amblyopone

Hypoponera

21 (20) Mandible long, strongly projecting below clypeal margin and with numerous, bidenticulate teeth that are found along the entire inner surface **(A)** . . *Amblyopone*

— Mandible short, closing tightly against clypeus and with only three teeth, grouped together near the tip; middle tooth smallest **(AA)**. . . . . *Prionopelta*⋆

Note: 1 sp. *P. antillana,* central Florida.

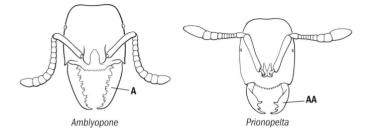

Amblyopone

Prionopelta

Note: Asterisks (⋆) denote introduced genera.

22 (20) Upper plate (tergite) of abdominal segment 4 coarsely and longitudinally costate **(A)**. Hind coxa with dorsal spine **(B)**. Tarsal claws always with a single subapical tooth, in addition to the terminal point **(C)**. (Ectatomminae) . . . . ***Gnamptogenys***

— Tergite of abdominal segment 4 lacking sculpture or with fine textured sculpture, never with striate or costate sculpture **(AA)**. Hind coxa lacking dorsal tooth or spine **(BB)**. Tarsal claws usually simple **(CC)** **23**

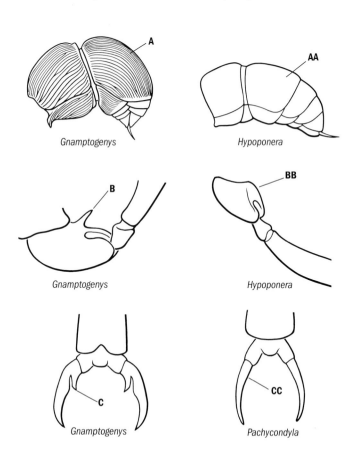

Gnamptogenys

Hypoponera

Gnamptogenys

Hypoponera

Gnamptogenys

Pachycondyla

23 (22) Dorsum of mesosoma lacking a promesonotal suture, with the metanotal suture usually lacking as well **(A)** (metanotal groove present in *Proceratium californicum*). Tergite of abdominal segment 4 always enlarged and strongly arched so that it forms the rearmost part of the abdomen when viewed in profile; apex of abdomen directed anteriorly **(B)**. (Proceratiinae) **24**

— Dorsum of mesosoma with at least the promesonotal suture; the metanotal suture usually present as well **(AA)**. Abdominal tergite 4 not strongly enlarged and only weakly arched; apex of abdomen directed rearwards or ventrally, not anteriorly **(BB)**. (Ponerinae) **25**

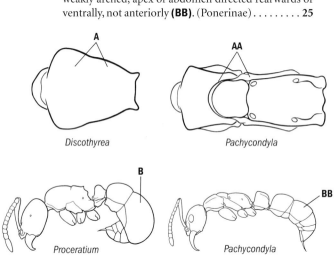

Discothyrea   Pachycondyla

Proceratium   Pachycondyla

24 (23) Apical antennal segment massive, about as long as the remaining segments combined (excluding scape) **(A)**. Mandible with only a single tooth at the tip **(B)**. Frontal carinae fused to form a single vertical ridge separating the antennal sockets **(C)**, which are located on a clypeal shelf that projects forward, covering the rear part of the mandibles when they are closed in full-face view. . . . . .
. . . . *Discothyrea*
Note: 1 sp. *D. testacea,* not commonly collected.

— Apical antennal segment moderately enlarged but distinctly shorter than the remaining segments combined (excluding scape) **(AA)**. Mandible with several distinct teeth **(BB)**. Frontal carinae separate and slightly expanded laterally and upwards from the plane of the head **(CC)**. Antennal sockets located at anterior margin of head; clypeal shelf absent. . . . . ***Proceratium***

Note: Not commonly collected.

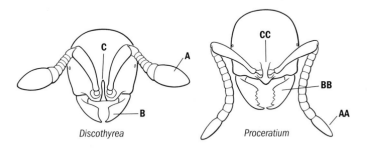

Discothyrea   Proceratium

---

25 (23) Mandible long and straight, in full-face view inserted in the middle of the front margin of the head, with two or three large teeth near tip; inner margin without teeth **(A)**. Petiolar node with at least one tooth or spine **(B)**. No visible constriction between abdominal segments 3 and 4 **(C)** **26**

— Mandible short to long and triangular to straight, in full-face view inserted at the side (lateral corner) of the head, with the teeth (when present) located along the inner margin **(AA)**. Petiolar node without spines or teeth **(BB)**. Visible constriction between abdominal segments 3 and 4 present **(CC)** **27**

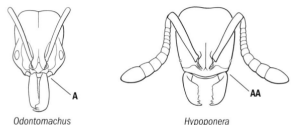

Odontomachus   Hypoponera

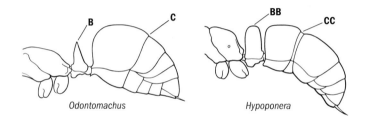

Odontomachus

Hypoponera

---

26 (25) Petiolar node with a prominent vertical tooth or spine **(A)**. With the head viewed from the back near the neck of the pronotum, the top of the head as in **(B)** . . . .
. . . . **Odontomachus**

— In side view, petiolar node strongly transversely compressed (narrow); and when seen from front or rear with dorsal margin concave, with a tooth on each lateral corner **(AA)**. With the head viewed from the back near the neck of the pronotum, the top of the head as in **(BB)**
. . . . **Anochetus***

Note: 1 sp. *A. mayri,* not commonly collected, central and southern Florida.

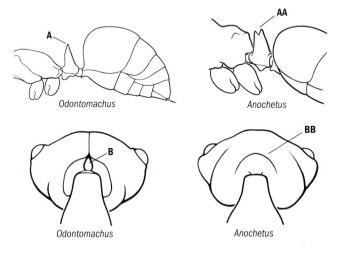

Odontomachus

Anochetus

Odontomachus

Anochetus

Note: Asterisks (*) denote introduced genera.

27 (25) Outer face of tibia of middle leg with many coarse, spine-like bristles in addition to hairs and apical spur **(A)**. Basal portion of mandible with a distinct dorsolateral oval pit near the insertion (difficult to see) **(B)** . . . .
. . . . ***Cryptopone***

> Note: 1 sp., *C. gilva*, not commonly collected,
> southeast and south central U.S.

— Outer face of tibia of middle leg may have hairs, but lacks coarse, spine-like bristles except at apex **(AA)**. Basal portion of mandible without a dorsolateral oval pit near the insertion **28**

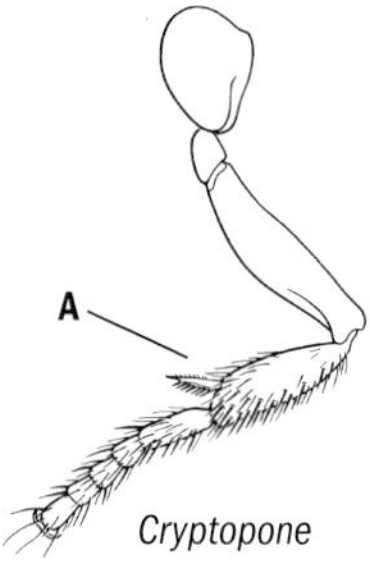

Cryptopone

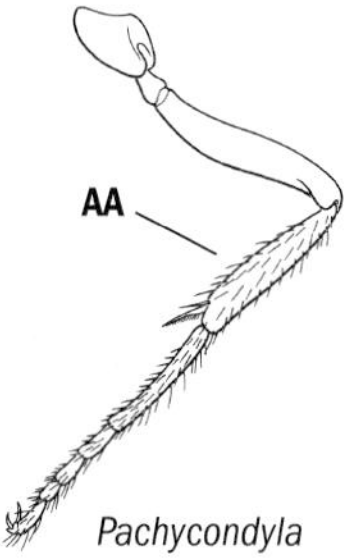

Pachycondyla

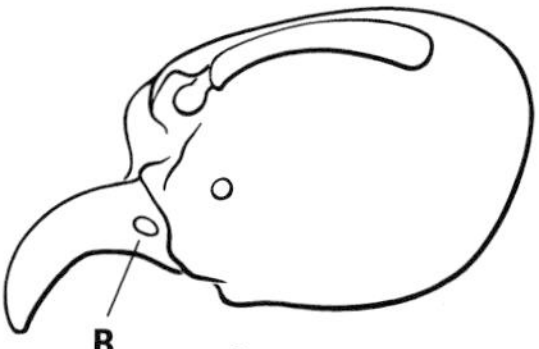

Cryptopone

28 (27) Small ants; total length 3 mm or less. The inner tip of hind tibia, when viewed from in front, with a single, large, comb-like (pectinate) spur; without a second smaller spur in front of the pectinate main spur **(A)**. **29**

— Large ants; total length greater than 4 mm. The inner tip of hind tibia, when viewed from in front, with either two pectinate spurs, one large and one small **(AA)**, or a large pectinate spur and a much smaller simple spur (use high magnification, 50 to 100 x) **(AAA)** **30**

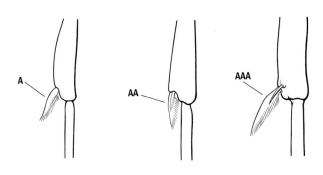

29 (28) Petiole, in side view, with lobe-like subpetiolar process that has a circular, often translucent thin spot (window) toward the front, and two small, sharp teeth or angles projecting posteriorly **(A)** . . . . *Ponera*

— Subpetiolar process a simple rounded lobe, without a translucent thin spot toward the front when viewed from the side and without sharp teeth or angles to the rear **(AA)**. . . . . *Hypoponera*

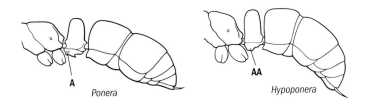

Ponera

Hypoponera

30 (28)   Promesonotal suture present and metanotal suture absent on mesosomal dorsum **(A)**. Tibia of middle and hind legs each with two pectinate spurs, one large and one small (teeth on the small spur can be difficult to see; use high magnification, 50 to 100 x) **(B)**. Frontal lobes and antennal sockets widely separated **(C)** . . . .
. . . . ***Platythyrea***

> Note: 1 sp. *P. punctata,* not commonly collected, southern Florida and extreme southern Texas.

—   Both promesonotal and metanotal sutures present on mesosomal dorsum **(AA)**. Tibia of middle and hind legs each with two spurs, one large and pectinate, and the other small and simple (use high magnification, 50 to 100 x) **(BB)**. Frontal lobes and the antennal sockets very close together **(CC)** **31**

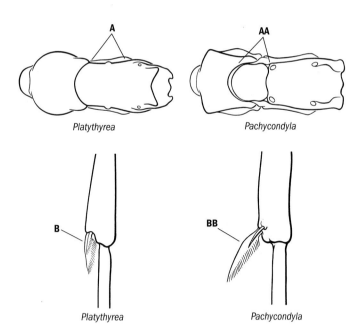

Platythyrea   Pachycondyla

Platythyrea   Pachycondyla

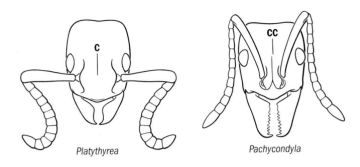

Platythyrea   Pachycondyla

---

31 (30) Tarsal claws on the hind legs finely pectinate **(A)**. Mandibles slender and elongated and without teeth on inner margin **(B)** . . . . ***Leptogenys***

— Tarsal claws on hind legs simple, without teeth on inner margin **(AA)**. Mandibles always subtriangular and armed with teeth **(BB)** . . . . **Pachycondyla**

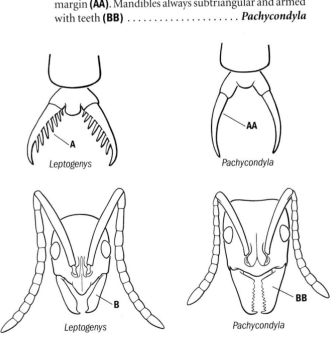

Leptogenys   Pachycondyla

Leptogenys   Pachycondyla

32 (3) Combining the following characters: First mesosomal segment freely articulating with the second mesosomal segment, forming a flexible joint **(A)**. Tibia of hind leg with a prominent, pectinate spur (use high magnification, 50 to 100 x) **(B)**. Eyes unusually large, often covering most of the sides of the head **(C)**. Ocelli present. (Pseudomyrmecinae) . . . . ***Pseudomyrmex***

— First mesosomal segment fused with the second mesosomal segment, forming an inflexible structure **(AA)**. Hind tibial spurs usually simple **(BB)**, sometimes pectinate **(BBB)** or absent. Eyes variously developed, sometimes vestigial or absent. When present, seldom unusually large, never covering most of the sides of the head. Ocelli nearly always absent **33**

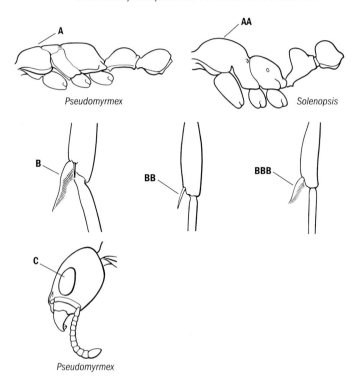

Pseudomyrmex

Solenopsis

Pseudomyrmex

33 (32) Antennal sockets completely exposed in full-face view, not covered by frontal lobes **(A)**; antennal sockets closely approximated, inserted at the anterior margin of the head **(A)**. Clypeus reduced. Eyes either a single facet or absent. (Ecitoninae) **34**

— In full-face view, antennal sockets covered at least in part by frontal lobes **(AA)**; antennal sockets usually well separated and set back from the anterior margin of the head at or near the posterior border of the clypeus **(AA)**. Clypeus developed. Eyes usually present, often with several facets or more **(BB)**. (Myrmicinae) **36**

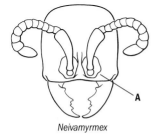

Neivamyrmex

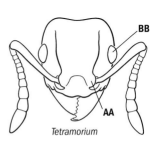

Tetramorium

34 (33) Tarsal claws simple, lacking a median tooth **(A)** . . . .
. . . . ***Neivamyrmex***

— Tarsal claws with a median tooth in addition to the terminal point **(AA)** **35**

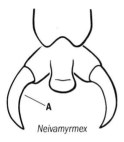

Neivamyrmex

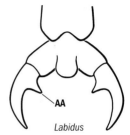

Labidus

35 (34) Apical width of antennal scape more than one third of its total length **(A)**. Propodeum in profile equipped with a pair of teeth or spines **(B)**. Workers weakly to moderately polymorphic. Majors lacking enormous heads. Entire body heavily sculptured and opaque . . . .
. . . . ***Nomamyrmex***

> Note: 1 sp. *N. esenbeckii* Wheeler, central and southern Texas.

— Apical width of antennal scape less than one third of its total length **(AA)**. Propodeum in profile without teeth or spines **(BB)**. Workers strongly polymorphic. Majors with enormous heads. The head (at least) relatively unsculptured and moderately to strongly shining. . . . . ***Labidus***

> Note: 1 sp. *L. coecus,* central and eastern Texas and adjacent states.

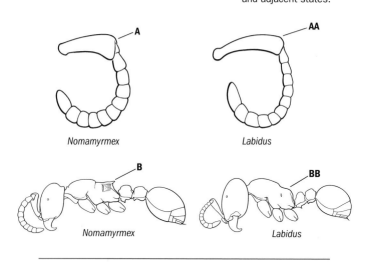

Nomamyrmex   Labidus

Nomamyrmex   Labidus

---

36 (33) Antenna with four or six segments, last two segments forming distinct club **(A)** **37**
— Antenna with seven segments, last two segments forming distinct club **(AA)** . . . . ***Eurhopalothrix***

> Note: 1 sp. *E. floridana,* not commonly collected, Florida.

— Antenna with 10 segments, last two segments forming distinct club **(AAA)** . . . . ***Solenopsis***

— Antenna with 11 segments, apical club variable **38**
— Antenna with 12 segments, apical club variable **(AAAA)**
. . . . **55**

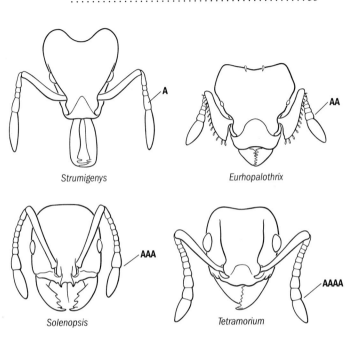

Strumigenys

Eurhopalothrix

Solenopsis

Tetramorium

37 (36) Mandibles thin, elongated, usually more or less straight, and with an apical fork consisting of two to three spine-like teeth at extreme tip **(A)**. Buccal cavity relatively long and narrow, lateral margins of cavity converging anteriorly and mandibles in ventral view apparently arising from apex of labral lobes **(B)**. . . . . ***Strumigenys***

— Mandibles short, often triangular or nearly so, with teeth or small denticles along inner margins **(AA)**, but never with a well-developed apical fork. Buccal cavity relatively short and wide, lateral margins of cavity not converging anteriorly and mandibles in ventral view arising to the outside of labral lobes **(BB)** . . . ***Pyramica***

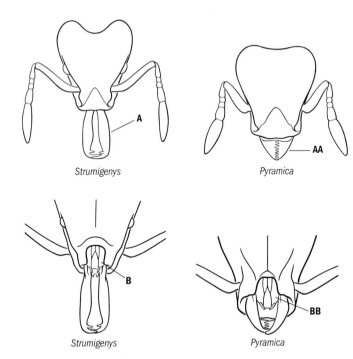

Strumigenys

Pyramica

Strumigenys

Pyramica

38 (36) Postpetiole attached to dorsal surface of abdominal segment 4 **(A)** and capable of flexing upwards over dorsal surface of body. Petiole strongly flattened and lacking dorsal node **(B)** .... ***Crematogaster***
— Postpetiole attached to anterior face of abdominal segment 4 **(AA)** and not capable of flexing upwards over dorsal surface of body. Petiole nearly always with distinct dorsal node **(BB)**. If node absent *(Xenomyrmex)*, petiole not strongly flattened. **39**

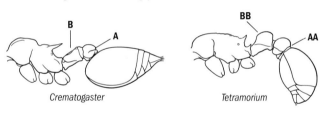

Crematogaster

Tetramorium

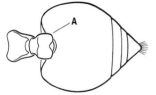

Crematogaster

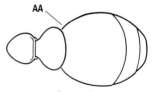

Tetramorium

---

39 (38) With head in side view, a sharply differentiated diagonal carina (ridge) present, extending up from mandible insertions to above the eye **(A)**. Antennal club often absent **(B)**. Head and body often with tubercles or spines **(C)** **40**

— With head in side view, lacking diagonal carina running from the mandible insertions to above the eye **(AA)**, though longitudinal rugae may be present on sides and front of head. Antenna often with a 2- or 3-segmented club **(BB)**. Head and body lacking tubercles, but spines or teeth may be present **(CC)** **44**

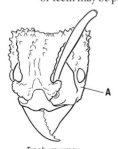

Trachymyrmex

Tetramorium

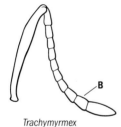

Trachymyrmex

Tetramorium

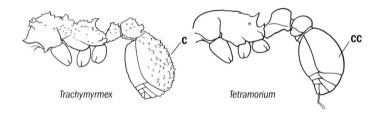

Trachymyrmex   Tetramorium

40 (39) Frontal lobes massively expanded laterally, in full-face view covering most of the sides of the head below the eye **(A)**. Head and body with flattened and appressed hairs; erect hairs absent. Tergite of abdominal segment 4 without tubercles **(B)** . . . . *Cyphomyrmex*

— Frontal lobes not expanded laterally to cover the sides of the head **(AA)**. Head and body with conspicuous erect hairs or setae, often stiff and curled. Abdominal tergite 4 sometimes tuberculate **(BB)** **41**

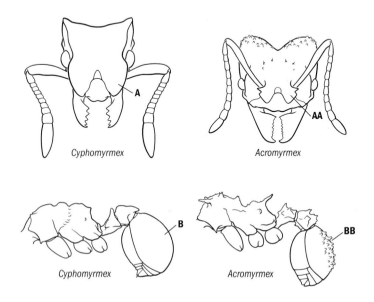

Cyphomyrmex   Acromyrmex

Cyphomyrmex   Acromyrmex

41 (40) Frontal lobes enlarged and reaching anterior margin of clypeus in full-face view **(A)**. Small, monomorphic species lacking spines, teeth, or prominent tubercles on mesosoma, except for two small teeth on the humeral angles, and two on the propodeum **(B)** . . *Mycetosoritis*
Note: 1 sp. *M. hartmanni,* central Texas to Louisiana.
— Frontal lobes variously shaped, but not reaching anterior margin of clypeus in full-face view **(AA)**. Conspicuous spines or tubercles present on mesosomal dorsum **(BB)**. Workers sometimes polymorphic **42**

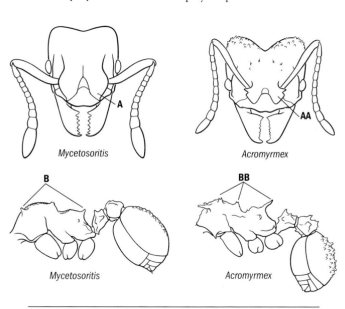

Mycetosoritis   *Acromyrmex*

Mycetosoritis   *Acromyrmex*

42 (41) Dorsum of mesosoma with three pairs of long spines **(A)** and without tubercles. Vertex and abdominal tergite 4 more or less smooth and without tubercles **(B)**. Strongly polymorphic . . . . *Atta*
— Dorsum of mesosoma with more than three pairs of long spines or with tubercles **(AA)**. Vertex and abdominal tergite 4 usually roughened and tuberculate **(BB)**. Monomorphic or moderately polymorphic. **43**

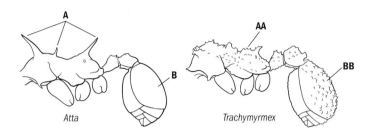

Atta

Trachymyrmex

---

43 (42) Mesosomal spines generally absent, tuberculate if present **(A)**. Frontal carinae well-developed, sometimes extending almost to the posterior corners of the head **(B)**. Entire head and mesosoma with at least some, often many, small tubercles **(C)**, which are blunt or sometimes bear short, sharp teeth. Monomorphic or very weakly polymorphic. . . . . ***Trachymyrmex***

— Mesosomal spines present, prominent, never tuberculate **(AA)**. Frontal carinae short, indistinct, ending well in front of the posterior corners of the head **(BB)**. On head and mesosoma, tubercles limited largely to the vertex and mesosomal dorsum, nearly absent elsewhere **(CC)**. Moderately polymorphic . . . . ***Acromyrmex***

Note: 1 sp. *A. versicolor,* western Texas to southern California.

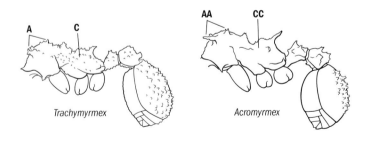

Trachymyrmex

Acromyrmex

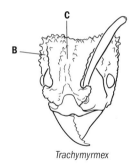

Trachymyrmex

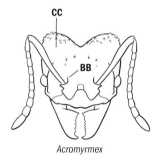

Acromyrmex

44 (39) Frontal carinae expanded to cover the sides of the head in full-face view **(A)**. Antennal scrobes deep, receiving entire scape **(A)**. Dorsum of mesosoma strongly flattened; promesonotum sharply marginate at side, often spinose **(B)**. Dimorphic or polymorphic . . . ***Cephalotes***

— Frontal carinae not greatly expanded to cover the sides of the head in full-face view **(AA)**. Antennal scrobes, when present, shallow and not receiving entire scape **(AA)**. Promesonotum not laterally marginate or spinose **(BB)**. Monomorphic, except for one rare, minute species **45**

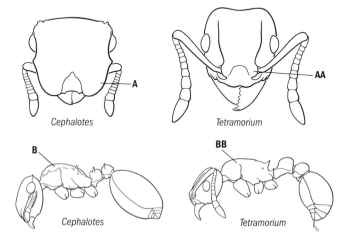

Cephalotes

Tetramorium

Cephalotes

Tetramorium

45 (44) Petiolar node absent or rudimentary **(A)** *and* propodeum rounded in profile without spines or teeth **(B)**
.... ***Xenomyrmex***

> Note: 1 sp. *X. floridanus,* southern Florida.

— Petiolar node nearly always present and well developed **(AA)**; if absent, then propodeum with spines or teeth **(BB)** 46

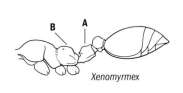

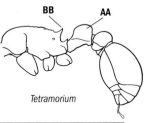

Xenomyrmex   *Tetramorium*

46 (45) Antennal club clearly 2-segmented **(A)** 47
— Antennal club 3-segmented **(AA)**, absent, or indistinct
.... 48

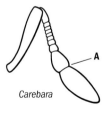

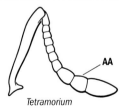

Carebara   *Tetramorium*

47 (46) Frontal carinae present, extending well past the eye, almost to the occipital margin **(A)**. Propodeum with a pair of long spines **(B)**. Eye with more than 10 facets **(C)**; lower margin of eye flat .... ***Wasmannia****

> Note: 1 sp. *W. auropunctata,* southern Florida.

— Frontal carinae short **(AA)**. Propodeum armed with a pair of small teeth **(BB)**. Eye minute, with five or fewer facets **(CC)**; lower margin of eye rounded. Dimorphic with major and minor workers; majors not yet recorded in North America .... ***Carebara***

> Note: 1 sp. *C. longii,* central Texas.

Note: Asterisks (*) denote introduced genera.

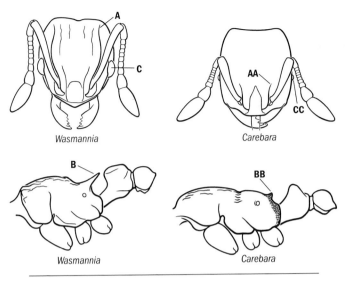

Wasmannia

Carebara

Wasmannia

Carebara

48 (46) Frontal carinae extending well past the eye, sometimes to the occipital margin **(A)**. **49**
— Frontal carinae ending far short of occipital margin, seldom surpassing the eye **(AA)** **51**

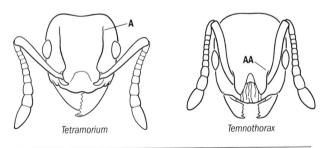

Tetramorium

Temnothorax

49 (48) Dorsum of head and mesosoma rugoreticulate. Mandible with six or seven teeth **(A)**. Area of clypeus immediately in front of antennal sockets raised up into a narrow ridge or shield wall so that sockets appear to be placed within deep pits **(B)**. Antennal scapes not flattened at base . . . . *Tetramorium* (in part)

— Mandible with either zero *or* four teeth **(AA)**. Area of clypeus immediately in front of antennal sockets flat and not raised up into a narrow ridge or shield wall; antennal sockets not appearing to be set within deep pits **(BB)**. Antennal scapes flattened at base. 50

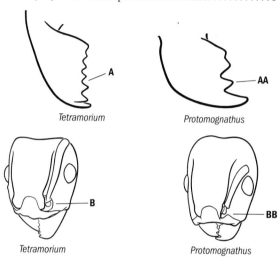

Tetramorium

Protomognathus

Tetramorium

Protomognathus

50 (48) Mandible with a broad cutting margin, but no discernible teeth. Anterior edge of clypeus with prominent median notch or deep concavity **(A)** . . . ***Harpagoxenus***
Note: 1 sp. *H. canadensis,* eastern Canada, northeastern U.S.

— Mandible with four teeth. Median impression of anterior margin of clypeus broad and shallow **(AA)** . . . .
. . . . ***Protomognathus***
Note: 1 sp. *P. americanus,* eastern North America.

Harpagoxenus

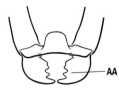

Protomognathus

51 (49) Eye rudimentary or absent. Mandible with four teeth and strongly oblique cutting margin. Propodeum rounded in profile **(A)** . . . . ***Dolopomyrmex***

> Note: 1 sp. *D. pilatus,* not commonly collected, known from southern Arizona, New Mexico, and California.

— Eye present, often with 10 or more facets. Mandible sub-triangular, with more than four teeth. Propodeum usually with teeth or spines **(AA)**, sometimes just angulate at juncture of dorsal and posterior faces. **52**

Dolopomyrmex   Temnothorax

---

52 (51) Eyes with short, erect hairs **(A)** (use high magnification)
. . . . ***Formicoxenus***
— Eyes lacking erect hairs. **53**

Formicoxenus

---

53 (52) Dorsal surface of petiole node armed with one or more pairs of short spines or tubercles **(A)** . . . . ***Nesomyrmex***

> Note: 1 sp. *N. wilda,* known from Rio Grande valley near Brownsville, Texas.

— Dorsal surface of petiole node unarmed, without a pair of short spines or tubercles **54**

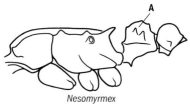

Nesomyrmex

---

54 (53) Mandible with six teeth **(A)**. Usually, median portion of clypeus smooth and longitudinally excavate, lacking carinae centrally; several carinae usually present on lateral portions **(B)** .... ***Leptothorax***

— Mandible with five teeth **(AA)**. Median portion of clypeus more or less flat, not smooth and longitudinally excavate, and with one or more carinae centrally, sometimes weakly developed or very rarely absent; carinae on lateral portions often present **(BB)** ....
.... ***Temnothorax*** (in part)

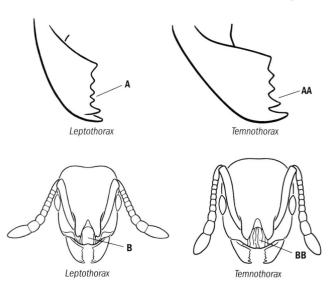

Leptothorax   *Temnothorax*

Leptothorax   *Temnothorax*

55 (36) Petiole short and subcylindrical to subquadrate, without anterior peduncle and with dorsal node absent or rudimentary **(A)**. Propodeum with two pairs of teeth: a short pair of teeth anterior to a longer pair on the propodeal angle **(B)**. In side view, the ventral margin of the head with a sharp longitudinal carina below the eye, extending from base of mandible to posterior corner of head **(C)**. Pronotal humeri angulate ...... ***Myrmecina***

— Petiole with distinct dorsal node **(AA)**. Propodeum unarmed or armed with a single pair of spines or denticles **(BB)**. Carina as described above absent **(CC)**. Pronotal humeri rounded, rarely subangulate............. **56**

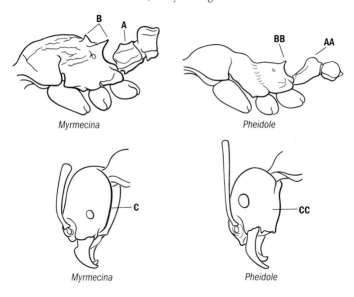

Myrmecina   Pheidole

Myrmecina   Pheidole

---

56 (55) Petiole with a large, anteroventral, plate-like process **(A)**. Anterior peduncle absent **(B)** .... ***Vollenhovia*** *
Note: 1 sp. *V. emeryi,* District of Columbia and adjacent Maryland.

Note: Asterisks (*) denote introduced genera.

— Plate-like, subpetiolar process as described above absent; a small anteroventral tooth may be present **(AA)**. Petiole often with a distinct peduncle **(BB)**; sometimes reduced or absent **(BBB)** 57

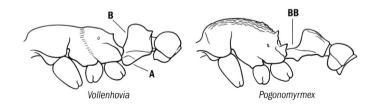

Vollenhovia

Pogonomyrmex

Tetramorium

---

57 (56) Area of clypeus immediately in front of antennal sockets raised up into a narrow ridge or shield wall so that sockets appear to be placed within deep pits **(A)**. Middle and hind tibial spurs simple or absent **(B)** (use high magnification, 50 to 100 x). In side view, dorsal node of petiole rounded or subquadrate **(C)** ....
.... ***Tetramorium*** (in part)

Note: *T. caespitum**, urban areas throughout temperate North America; *T. tsushimae**, around St. Louis, Missouri.

— Area of clypeus immediately in front of antennal sockets usually flat and not raised up into a narrow ridge or shield wall; thus, antennal sockets do not appear to be set within deep pits **(AA)**. If narrow ridge appears to be present **(AAA)**, then middle and hind tibial spurs pectinate (use high magnification, 50 to 100 x) **(BB)** and dorsal node of petiole more or less triangular in side view **(CC)** 58

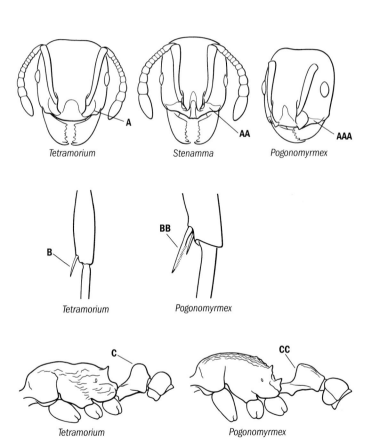

Tetramorium   A

Stenamma   AA

Pogonomyrmex   AAA

B   Tetramorium

BB   Pogonomyrmex

C   Tetramorium

CC   Pogonomyrmex

58 (57) Combining the following: In dorsal view, postpetiole subcircular *and* notably swollen relative to the petiole **(A)**. Dorsum of head and mesosoma very finely sculptured, dull and lacking erect hairs. Clypeus with long median seta that projects forward over the mandibles **(B)** .... *Cardiocondyla**

Note: Asterisks (*) denote introduced genera.

— Not as above: postpetiole variable in shape and size, but not *both* subcircular and notably swollen relative to the petiole in dorsal view **(AA)**. Sculpture on dorsum of head and/or mesosoma, when present, often including rugae or punctures; erect hairs usually present on dorsum of head and mesosoma. Median clypeal seta sometimes absent **(BB)** **59**

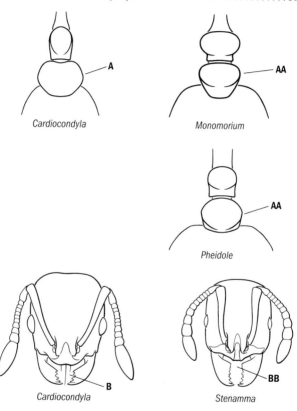

Cardiocondyla

Monomorium

Pheidole

Cardiocondyla

Stenamma

59 (58) Psammophore (long, curved hairs on underside of head) present **(A)**. Mesosoma in side view weakly to moderately convex; metanotal impression absent or indistinct **(B)**. Petiolar peduncle well developed, mostly smooth and strongly shiny. Petiolar node usually has a short, often nearly vertical anterior face that meets the long, gradually sloping posterior face to form a sharp (sometimes rounded) peak, giving the node a triangular shape overall **(C)**. . . . . ***Pogonomyrmex*** (in part)

— Psammophore generally absent. Mesosomal profile variable, but if psammophore present, then propodeum depressed below level of mesonotum in side view **(BB)**. Other characters variable. **60**

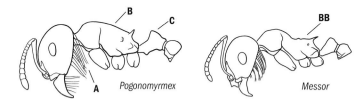

Pogonomyrmex

Messor

---

60 (59) Combining the following: Anteromedian portion of clypeus notably elevated and bicarinate (carinae usually ending to form two submedian teeth on the anterior clypeal margin) **(A)**. Propodeum rounded or rarely angular, unarmed, without spines or teeth **(B)**. Antennal club 3-segmented **(C)**. . . . . ***Monomorium***

— Clypeus generally not bicarinate; if so **(AA)**, propodeum usually armed with teeth or short spines **(BB)**. Antennal club variable **(CC)**, sometimes absent **61**

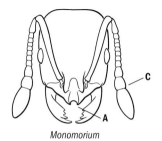

Monomorium

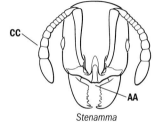

Stenamma

Monomorium   *Stenamma*

61 (60) Propodeum depressed below level of promesonotum in
side view **(A)** *and/or* metanotal impression substantial
(B), separating promesonotum and propodeum into
separate, evenly rounded convexities **62**

— Propodeum not depressed below level of promesonotum **(AA)**. Metanotal impression often absent, or slight
when present, just briefly interrupting the mesosomal
profile **68**

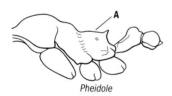

Pheidole

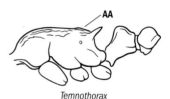

Temnothorax

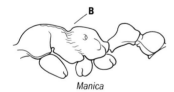

Manica

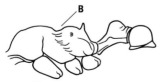

Temnothorax pergandei

62 (61) Propodeum slightly depressed below level of promesonotum **(A)** *and* antennal sockets widely separated by the
median posterior portion of the clypeus, which forms a
broadly rounded or triangular shape **(B)**. Metanotal impression sometimes substantial **(C)**, separating promesonotum and propodeum into separate, evenly
rounded convexities **63**
— Propodeum usually notably depressed below level of
promesonotum **(AA)**; if only slightly depressed, then antennal sockets closely approximated, with the median
posterior portion of the clypeus forming a narrow,
finger-like lobe **(BB)**. Metanotal impression variable.
Dorsal surface of propodeum often flattened **(AA)**, but
sometimes convex **65**

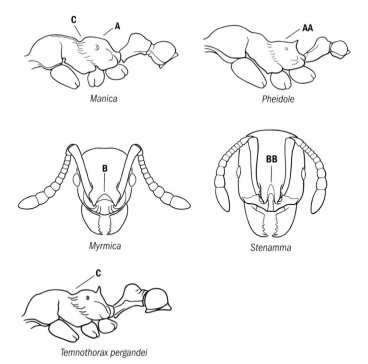

Manica

Pheidole

Myrmica

Stenamma

Temnothorax pergandei

63 (62) Mandible with five teeth **(A)**. Middle and hind tibial spurs simple **(B)**. Propodeum armed with short teeth **(C)** .... *Temnothorax pergandei*

— Mandible with seven or more teeth **(AA)**. Middle and hind tibial spurs pectinate (use high magnification, 50 to 100 x) **(BB)**. Propodeum with teeth or spines **(CC)** or unarmed **(CCC)**............................. **64**

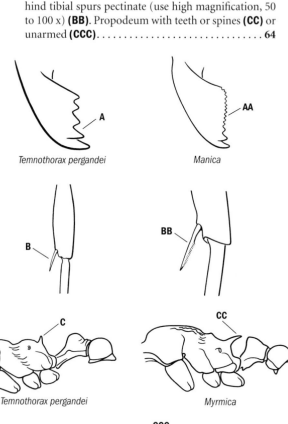

Temnothorax pergandei

Manica

Temnothorax pergandei

Myrmica

Manica

64 (63) Propodeum unarmed **(A)** .... *Manica*
— Propodeum with teeth or spines **(AA)** ....
.... *Myrmica* (in part)

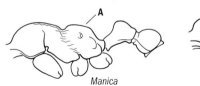
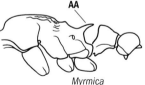

Manica   Myrmica

---

65 (62) Antennal sockets closely approximated; thus, median
posterior portion of clypeus forms a narrow, finger-like
projection that extends rearwards between the frontal
lobes **(A)**. Anterior median portion of clypeus often
with a pair of fine, longitudinal carinae that diverge anteriorly **(B)**.................... *Stenamma* (in part)
— Antennal sockets not closely approximated **(AA)**, the
median posterior portion of the clypeus forming a triangular or broadly rounded shape extending rearwards
between the frontal lobes. Clypeus never bicarinate **(BB)**
.... **66**

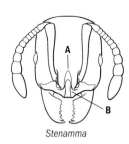
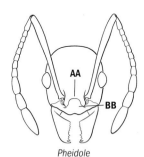

Stenamma   Pheidole

66 (65) Antenna with a distinct 3- or (rarely) 4-segmented apical club **(A)**. Palp count 3,2 or 2,2. Worker caste dimorphic, rarely strongly polymorphic . . . . ***Pheidole***

— Antenna without distinct apical club **(AA)**. Palp count 5,3 or 4,3. Workers monomorphic or weakly to moderately polymorphic **67**

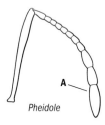
Pheidole

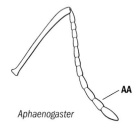
Aphaenogaster

67 (66) Head quadrate **(A)**. Mandible short and thick, with outer margin strongly curving toward the midline **(B)**. Psammophore often present . . . . ***Messor***

— Head longer than broad, often noticeably narrowed toward the vertex **(AA)**. Mandible slender and triangular, with outer margin not strongly curving toward midline **(BB)**. Psammophore absent . . . . ***Aphaenogaster***

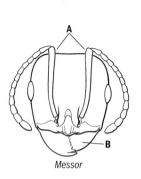
Messor

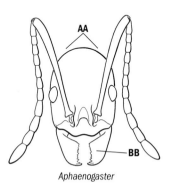
Aphaenogaster

68 (61) Metanotal impression small but present, interrupting the mesosomal profile in side view as a modest notch or indentation **(A)** **69**

— Metanotal impression absent **(AA)**; mesosomal profile without interruption, evenly convex or flattened, or very slightly undulating **70**

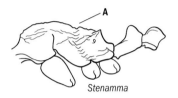

Stenamma

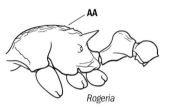

Rogeria

69 (68) Antennal sockets closely approximated; median posterior portion of clypeus forming a narrow, finger-like lobe extending rearwards between the two frontal lobes **(A)**. Tibial spur simple **(B)**. Antennal scape gradually and evenly bent as it approaches the insertion **(C)** .....
.... ***Stenamma*** (in part)

— Antennal sockets not closely approximated; median posterior portion of clypeus forming a triangular or broadly rounded shape extending rearwards between the frontal lobes **(AA)**. Tibial spur pectinate **(BB)**. Antennal scape often sharply bent near insertion; the bend frequently features a conspicuous ridge or lamina **(CC)**
.... ***Myrmica*** (in part)

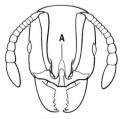

Stenamma

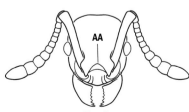

Myrmica

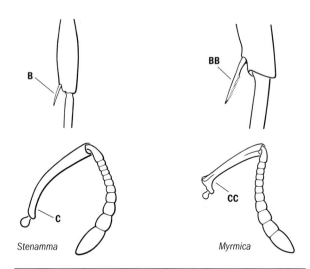

Stenamma

Myrmica

70 (68) Antennal sockets closely approximated; median posterior portion of clypeus forming a narrow, finger-like lobe extending rearwards between the two frontal lobes **(A)**. Clypeus bicarinate, with the carinae often small (use high magnification), and sometimes not extending to the anterior margin **(B)**. Eyes very small (less than 15 facets, usually less than 10) . . . . ***Rogeria***

— Antennal sockets not closely approximated; median portion of clypeus forming a triangular or broadly rounded shape extending rearwards between the frontal lobes **(AA)**. Clypeus never bicarinate **(BB)**. Eyes well developed, generally moderate to large in size (more than 15 facets) **71**

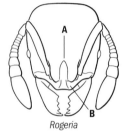

Rogeria

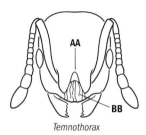

Temnothorax

71 (70) Mandible with more than five teeth **(A)**. Antennal club absent **(B)**. Head and mesosomal dorsum covered with dense rugoreticulate sculpture **(C)** .... ***Pogonomyrmex*** (in part)

— Mandible with five teeth **(AA)**. Antennal club 3-segmented **(BB)**. Sculpture on head and mesosoma variable, seldom rugo-reticulate on both head and mesosomal dorsum **(CC)** .... ***Temnothorax*** (in part)

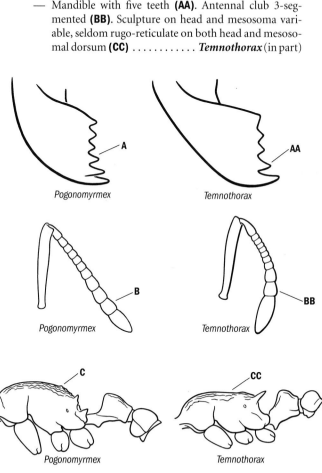

Pogonomyrmex

Temnothorax

Pogonomyrmex

Temnothorax

Pogonomyrmex

Temnothorax

TAXONOMIC DESCRIPTIONS

Subfamily Descriptions

Subfamily Amblyoponinae

The subfamily Amblyoponinae has a worldwide distribution and is comprised of nine genera, with only *Amblyopone* (four species) and one introduced species of *Prionopelta* present in North America. The members of this subfamily are mostly cryptobiotic predators that nest in soil or ground litter.

Members of the genus *Amblyopone* are specialist predators of certain arthropods, especially centipedes, living in soil or rotten wood. The bizarre feeding habits of *Amblyopone* extend well beyond their prey specialization and have led some to call this group by the evocative name "dracula ants." Some species of *Amblyopone* practice a form of nonlethal cannibalism in which the queens obtain nearly all of their nutrients from the hemolymph of larvae, whose integument they unceremoniously puncture prior to feeding. This behavior, called larval hemolymph feeding (LHF), has also been observed in *A. oregonensis* by Alex Wild in California's Sierra Nevada. Despite these specialized feeding habits, the ants of this genus also possess some behavioral and morphological traits considered "primitive" (ancestral) for ants as a whole. For example, the abdominal segment 2 (petiole) is broadly attached to segment 3 and lacking a distinct posterior face.

Subfamily Cerapachyinae

The Cerapachyinae are largely tropical and best represented in the Old World, especially in the Australasian and Malagasy regions. The biology of most Cerapachyines is little known, but many species appear to be specialized predators (feeding on the brood of other ants, or on termites) and some may have seminomadic life-histories. In addition, many species have wingless, worker-like queens. Several rare species occur in the southwestern United States, from Arkansas to California.

Subfamily Dolichoderinae

This relatively small subfamily is most diverse and abundant in tropics of both the Old and New Worlds. The few genera present in North America are at the northern limits of their ranges. Seven genera occur in our area; two of these are introduced. Most of our species are primarily southern in their distributions. One, *Tapinoma sessile,* is omnipresent from southern Canada south into Mexico.

The dolichoderines lack a sting, but instead are armed with defensive compounds. These chemicals are produced by the anal gland, a structure unique to the subfamily. This gland is the source of the pungent and often unpleasant odors produced by many species when they are disturbed, crushed, or otherwise annoyed. In many species, colonies may be polygynous and/or polydomous. Some are predaceous, but most appear to be generalized scavengers with a strong liking for sugars taken directly or indirectly from plant sources. In addition to founding colonies via single, newly mated queens, some produce new colonies by budding or fission. Most dolichoderines are free-living; a few are social parasites on other members of their subfamily.

Subfamily Ecitoninae

These are the New World legionary, or army, ants, which were formerly placed in the subfamily Dorylinae with the very different Old World forms. About two dozen species, mostly in the genus *Neivamyrmex,* occur across the southern United States.

Army ants are among our most interesting ants. They have no permanent nests; rather, the entire colony regularly migrates to new sites. The activities of the colony are determined by the egg-laying cycle of the queen and the developmental cycle of the brood, both of which are strongly periodic in these ants. During the statary phase, the colony establishes a temporary nest, or "bivouac," within a hollow log or other suitable cavity and remains there for about three weeks. During this time, the queen becomes physogastric (i.e., develops a highly swollen gaster) and lays an enormous batch of eggs. At the same time, the larvae of the previous egg batch have completed development and spend the statary phase as pupae. Columns of foragers leave the bivouac to collect food, usually at night or during cloudy or rainy weather, events that are called "raids." They commonly attack nearby nests of other ants, but only prey on insects and other arthropods on an opportunistic basis. Even small vertebrates may occasionally fall prey to these ants.

The statary phase ends when, simultaneously, the new eggs hatch into hungry larvae and the pupae eclose (emerge) to produce a new generation of workers. This greatly increases the colony's demand for food and commences the nomadic phase. During this time, every night begins with a strong raid that turns into an emigration—the entire colony moves to a new temporary nest. As the larvae approach maturity, the emigration raids diminish in intensity. Once most of the larvae have pupated, the queen again becomes physogastric and the colony re-enters the statary phase.

Army ant queens are never winged. Males are much larger than the workers and are powerful fliers, often attracted to lights at night. Many of our species are almost exclusively subterranean foragers and are therefore collected very infrequently. A persistent and annoying problem in army ant taxonomy is that many taxa were described based on males not associated with workers. Thus, two parallel classifications arose, one based on males and the other on workers. There has been slow progress toward unity, but the process has been hampered by the fact that males are seldom taken inside nests. It will be many years before all of these male and worker-based "species" are resolved into a single classification.

Subfamily Ectatomminae

This subfamily includes four genera largely confined to tropical and warm temperate climates of New World, Old World and Indo-Australian regions. Some species avidly collect plant sap and plant nectary secretions. This group is represented by only two *Gnamptogenys* species (one introduced) in the extreme southern U.S.

Subfamily Formicinae

The subfamily Formicinae has worldwide distribution and is second only to the Myrmicinae in numbers of species. It is dominated globally by the very large and complex cosmopolitan genus *Camponotus*. In North America, the genera *Formica* and *Lasius* are most abundant and ecologically important. In total, 11 genera occur in our area, two of which are introduced from other regions. Most Formicines are trophic generalists, and many derive much of their diet from nectar and other plant exudates, directly or indirectly via Hemipterans. Most species nest in soil or in dead or rotting wood on or near the soil. In addition, a substantial minority of truly arboreal species exists, most belonging to the genus *Camponotus.* The taxonomy of most of the North American groups seems to be fairly well worked out. A major exception is the genus *Formica,* particularly those species comprising the *rufa* group. Note: the tropical Asian genus *Plagiolepis* has been recorded in California, but is not included in the key because it is now considered extirpated.

Subfamily Myrmicinae

This is by far the largest subfamily of ants, with approximately 150 recognized genera worldwide. Of these, 35 are known to occur in North America; seven of these are introduced. This subfamily also includes some of the largest ant genera, such as *Crematogaster, Pheidole, Tetramorium, Monomorium,* and *Solenopsis.* Myrmicines are unique in many respects. While they retain the classic hymenopteran sting found in ponerines and aculeate wasps, it has been modified to serve other functions in many myrmicines. In some genera, ants not only employ the sting as a weapon, but also use it for trail-laying and territorial marking. In others (e.g., *Pheidole*), the sting has become specialized for those latter purposes and is no longer used as a piercing weapon. In still others (e.g., *Crematogaster*), it has become an "applicator," with a flattened, blade-like portion that's used to smear poison gland secretions on the integument or skin of enemies or prey.

Ecologically, perhaps the most important myrmicine evolutionary innovation is that they are the only group of ants to make extensive use of starches as a food source. This has led to the evolution of seed-harvesting as a major life-history feature; the ants collect seeds and use them as storable food sources, particularly in arid climates. Another remarkable innovation is fungus-growing, present only in the myrmicine tribe Attini. These ants cultivate fungi by processing plant matter either themselves or with the help of caterpillars; the ants collect the caterpillar droppings and bring them into the nest. Overall, myrmicines exhibit most of the life-history innovations that ants have evolved, including all types of social parasitism, brood theft, specialized predation, pleometrosis, complex caste systems, polygyny, and polydomy.

Subfamily Ponerinae

Recently Bolton (2003) recognized the need to divide the traditional subfamily Ponerinae into six distinct subfamilies: Amblyoponinae, Ectatomminae, Heteroponerinae, Paraponerinae, Ponerinae, and Proceratiinae. Molecular studies strongly support the need for the breakup and agree with the subfamily divisions.

Ponerinae, as presently constituted, constitutes the largest subfamily of the six and includes 23 genera. Distribution is mainly tropical, but extends into subtropical (and even temperate regions) on all major continents. In North America, eight genera are present and include roughly 20 species. These ants vary widely in their life-histories and ecologies, but colonies are generally small and the ants are mostly predaceous.

Subfamily Proceratiinae

The subfamily Proceratiinae contains three genera that have a worldwide distribution. Most proceratines are tropical or subtropical, but a handful are found in the temperate zone. Observations of these ants indicate they are specialist predators of various arthropod eggs. In North America, two genera are present, *Discothyrea* (one species) and *Proceratium* (eight species). These ants are rarely collected because both the ants and their colonies are small and the workers seldom appear above ground.

Subfamily Pseudomyrmecinae

This pantropical subfamily consists of only three genera, but two of them, *Pseudomyrmex* in the New World and *Tetraponera* in the Old World, are diverse and abundant in most tropical woodlands and forests. The third genus, *Myrcidris,* is monotypic and occurs in Brazil.

These ants are predominantly arboreal, though a few species nest in dead wood on the ground, and one or two are even found in termite nests. The New World tropics are home to perhaps 150 to 200 *Pseudomyrmex* species. Some are obligate inhabitants of specialized ant plants, and a very few are social parasites, but the vast majority are free-living. Ten species of *Pseudomyrmex* occur in the southern United States.

Genus Descriptions

ACANTHOSTICHUS Cerapachyinae
3 NA spp.

DIAGNOSTIC REMARKS: Most likely to be confused with *Cerapachys*, it can be distinguished by its 12-segmented antennae (11 in *Cerapachys*), the strongly flattened antennal scape, and the lack of a prominent lateral carina bordering the antennal sockets.

DISTRIBUTION AND ECOLOGY: *Acanthostichus* is a neotropical genus that reaches the northern limit of its distribution in the southwestern U.S., where two extremely rare species occur. As these

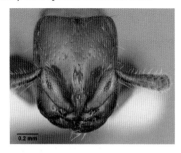

ants are suspected of being deeply subterranean, little is known of their habits. Specimens have been found under stones and other objects providing ground shelter, and collections have been made in association with nests of termites. MacKay (1996) is the essential reference.

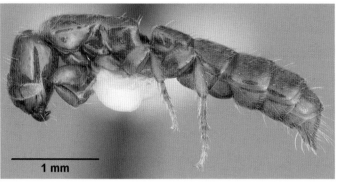

Acanthostichus punctiscapus

ACROMYRMEX   **Myrmicinae**
1 NA sp.

DIAGNOSTIC REMARKS: The genus *Acromyrmex* occupies a morphological middle ground between the highly polymorphic genus *Atta* and the monomorphic genus *Trachymyrmex*. *Acromyrmex* is separated from *Atta* by the presence of three pairs of spines or teeth on the promesonotum (*Atta* has two pairs), and from *Trachymyrmex* by the presence of spines or teeth on the promesonotum (rather than tubercles), by the absence of shallow antennal scrobes (often present in *Trachymyrmex*), and by its polymorphic worker caste (monomorphic in *Trachymyrmex*).

DISTRIBUTION AND ECOLOGY: Only one species *(A. versicolor)* of this

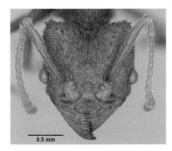

primarily neotropical group is found in the U.S., sporadically from western Texas to southern California and commonly in the Sonoran desert of southern Arizona. Like all attini, *Acromyrmex* species cultivate fungus, and like *Atta* they are true leafcutters. Most *Acromyrmex* inhabit tropical forests or grass-

Acromyrmex versicolor

lands. *A. versicolor* is atypical because it is a true desert ant. It makes large nests with multiple craters, and colonies are often founded by groups of newly mated queens (pleometrosis).

ACROPYGA   **Formicinae**
1 NA sp.

DIAGNOSTIC REMARKS: Unique in the context of the North American ant fauna because it is the only formicine genus in which the workers have vestigial eyes.

DISTRIBUTION AND ECOLOGY: This relatively small genus occurs in both Old and New World tropics. The ants are exclusively subterranean and have close mutualistic relationships with root-feeding coccids and aphids. One rare species (*A. epedana*) occurs sporadically at mid-elevations in southern Arizona and probably in northern Mexico. Lapolla's (2004) New World revision sheds light where there was only darkness before.

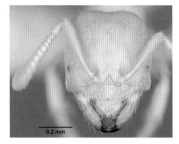

Acropega epedana

AMBLYOPONE **Amblyoponinae**
4 NA spp.

DIAGNOSTIC REMARKS: Easily recognized by the long, coarsely toothed falcate mandibles, the numerous, peglike teeth on the anterior border of the clypeus, and the broad attachment of petiole to succeeding abdominal segments. The related genus *Prionopelta* has mandibles with only three teeth.

DISTRIBUTION AND ECOLOGY: North American *Amblyopone* make diffuse, cryptic nests in soil, litter, and rotten wood. *A. pallipes* is found from eastern Canada south to Florida, and west to parts of

California. A forest species, this ant is occasionally collected in open habitats as well. Wherever it occurs, *A. pallipes* is seldom seen because the workers are subterranean foragers. It is normally a specialized predator of centipedes, but in the eastern U.S., workers will come to the surface of forest litter on warm, wet spring nights to harvest inchworms (Geometrid caterpillars). Colonies are small and often

Amblyopone pallipes

polygynous. *A. oregonensis* is the ecological analog of *A. pallipes* in northern California and the Pacific Northwest. *A. trigonig-natha* is known only from the holotype, collected in North Carolina. Ward (1988) contains a key that separates these species. A very different species—the minute, yellow, subtropical *A. oriza-bana*—was collected recently (2003) in southern Arizona by Markus Ruger (Fisher and Cover, unpublished collection data).

ANERGATES
1 NA sp.

Myrmicinae

`INTRODUCED`

DIAGNOSTIC REMARKS: A workerless inquiline impossible to confuse with any other North American ant. The males are found only in nests of the host and are pupoidal—cream to yellow in color, wingless, and barely able to walk. The queens are tiny and physogastric when found in the host nest. When found outside a host nest, they are dispersing, and may be recognized by the unique, prominent median longitudinal depression on the dorsal surface of the gaster.

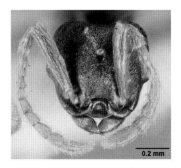

DISTRIBUTION AND ECOLOGY: *Anergates atratulus* is a workerless social parasite in the nests of *Tetramorium caespitum*. *Anergates* is native to Europe, but

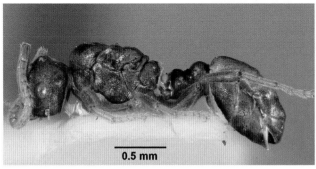

Anergates atratulus

also known from a handful of records in the eastern U.S., from Washington, D.C., to Connecticut. It was almost certainly imported along with its host, perhaps during colonial times. *Anergates*-infested colonies typically lack a host queen and produce nothing but males and females of the parasite. It is not known what happens to the host queen when an *Anergates* female invades the nest. She could be killed or expelled from the nest by the host workers, as the parasite queen seems far too puny to dispose of her unaided. Ray Sanwald (personal communication), a sagacious observer of ants on Long Island, New York, believes that *Anergates* females can gain entrance only into queenless host colonies. He says that the best way to collect *Anergates* is to remove the queen from a *T. caespitum* colony in the spring and come back the following year! There might be something to this . . .

ANOCHETUS   Ponerinae
1 NA sp.   **INTRODUCED**

DIAGNOSTIC REMARKS: Recognized by the unique head shape and mandibular structure. It can be confused with *Odontomachus,* but *Anochetus* has two teeth at the apex of the petiolar node (vs. one tooth or spine in *Odontomachus*). It might also be mistaken for *Strumigenys,* but its 1-segmented waist distinguishes it from

0.5 mm

Anochetus mayri

the 2-segmented *Strumigenys*. See Brown (1978) for a world revision of the genus.

DISTRIBUTION AND ECOLOGY: Our only species, *A. mayri*, has a circum-Caribbean distribution. It has been found in litter several times in central and southern Florida. The workers are ~2.5 mm long and have minuscule eyes. They are notably smaller than the workers of sympatric *Odontomachus* species, which are usually 6 to 9 mm long.

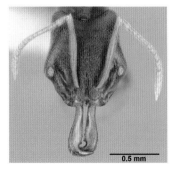

Anochetus mayri

APHAENOGASTER   **Myrmicinae**
<40 NA spp.

DIAGNOSTIC REMARKS: Small *Aphaenogaster* workers are sometimes confused with minor workers of *Pheidole*, but may be distinguished by the lack of a 3-segmented antennal club and by a 5,3 palp formula (as opposed to 2,2 or 3,2 in *Pheidole*). The worker caste in *Aphaenogaster* is monomorphic or weakly poly-

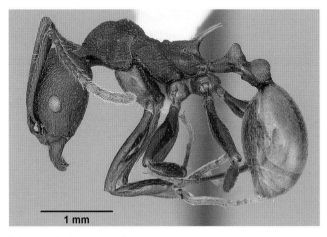

Aphaenogaster tennesseensis

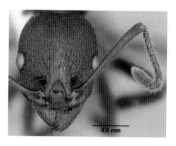

Aphaenogaster tennesseensis

0.5 mm

morphic; in *Pheidole,* the worker caste is dimorphic or strongly polymorphic. The taxonomy of North American *Aphaenogaster* is complex and only partly resolved. The major impediment to progress is the *rudis* complex: a large cluster of poorly differentiated sibling species that abound in the eastern half of the continent. The study of this complex is far from complete, but a valuable paper by Umphrey (1996) has made the nature of the problem clear. There are also minor taxonomic tangles to resolve among the western species.

DISTRIBUTION AND ECOLOGY: *Aphaenogaster* is an ecologically diverse group that is common throughout much of North America. Many of the *rudis* group forms are associated with eastern deciduous forest, where they are often the dominant ants on the forest floor. In these forest habitats, many spring ephemeral herbaceous plants depend on *rudis* group *Aphaenogaster* for seed dispersal. Some *Aphaenogaster* occur in open, grassy habitats, pine barrens, and sandhill vegetation types, while a couple of species are arboreal and a handful are true xerophiles. All are generalist scavengers and predators, with slender, long-legged workers. *Aphaenogaster* usually build colonies in soil or under rocks, but in forest habitats these ants may nest in rotten logs, branches, stumps, and occasionally live trees. In the desert southwest, *A. albisetosa* and *A. cockerelli* are abundant at low to mid-altitudes in arid habitats. Both species form conspicuous nests with sloppy gravel craters. The large workers are active diurnal foragers, especially on cloudy or wet days.

ATTA   Myrmicinae
2 NA spp.

DIAGNOSTIC REMARKS: *Atta* may be distinguished from the closely related genus *Acromyrmex* by the presence of three pairs of spines on the promesonotum (two pairs in *Acromyrmex*). The genus is desperately in need of revision, and it is simply astonishing that it remains such a taxonomic mess, given the economic importance of the ants and the generous funding that has been provided to study them. Fortunately, this deplorable situation does not impede recognition of the two *Atta* that occur in our area.

DISTRIBUTION AND ECOLOGY: This attine genus is a dominant group in the neotropics, and the ants are significant agricultural, forestry, and horticultural pests. Only two species are found in the United States: one *(A. texana)* in central and eastern Texas and Louisiana, and another *(A. mexicana)* in extreme southern Arizona at Organ Pipe Cactus National Monument. Like all attines, they are fungus growers. They form enormous colonies that forage widely and cut large quantities of green vegetation to serve as growing media for their fungus cultures. The workers are highly polymorphic and exhibit a well-developed division of labor. *Atta* species have adapted to life in diverse habitats. While most occur in tropical forest or savanna habitats, *A. texana* has become adapted to life in warm temperate forest gaps, and *A. mexicana* is a true xerophile.

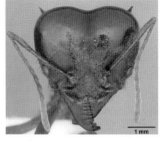

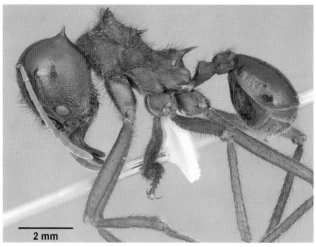

Atta mexicana

BRACHYMYRMEX Formicinae
6–10 NA spp.?

DIAGNOSTIC REMARKS: *Brachymyrmex* are most easily confused with small *Paratrechina*, but they are easily distinguished by their 9-segmented antennae (vs. 12 in *Paratrechina*). The pantropical tramp *Plagiolepis alluaudi*, reported from California, also resembles a small yellow *Brachymyrmex*, but it has 11 antennal segments.

DISTRIBUTION AND ECOLOGY: This predominantly neotropical group exists in a state of taxonomic chaos. Perhaps six to 10 species occur in the U.S., but they have never been defined and described adequately. Currently, the species name *depilis* is a waste-

basket for any of the small, subterranean yellow forms found from southern Canada to the Gulf Coast and south into Mexico. Several brown to gray species exist along the Gulf Coast; *B. obscurior* is the name used for all of them. Some of these ants may be exotics, as is one distinct form (incorrectly called *B. "musculus"* but now known to be

Brachymyrmex patagonicus

B. patagonicus) that is now spreading along the Gulf Coast in a variety of habitats. Workers gather secretions from root-feeding aphids and coccids in addition to being scavengers. In a fit of bad temper, Creighton (1950) referred to *Brachymyrmex* as a "miserable little genus," which may partly account for the lack of interest in the beasts shown by subsequent myrmecologists.

CAMPONOTUS   Formicinae

>50 NA spp.

DIAGNOSTIC REMARKS: Set apart from other formicines in North America by the absence of a ring of hairs around the acidopore, by the fact that the antennal sockets are set well back from the posterior border of the clypeus (instead of positioned right at the posterior border, as in other formicines), and by the lack of a metapleural gland. The worker caste is usually polymorphic, but sometimes dimorphic.

DISTRIBUTION AND ECOLOGY: This is an enormous, complex genus with a worldwide distribution. In the U.S. and Canada, there are numerous species, the majority of which belong to several reasonably well-defined species groups still referred to by their old subgeneric names.

---

1) Subgenus *Camponotus*

These are large, omnivorous ants that live in forested habitats and dwell in standing and fallen dead wood. They are abundant in cool temperate and boreal regions, where they play important roles in the wood-breakdown cycle of forest ecosystems. Some species (e.g., *C. pennsylvanicus* in the east, *C. modoc* in the west) are notable pests because they cause structural damage to damp wood in human dwellings.

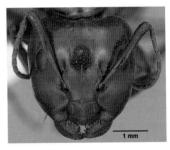

Camponotus (Camponotus) schaefferi

Camponotus (Camponotus) schaefferi

2) Subgenus *Tanaemyrmex*

These relatively large, soil-dwelling ants are especially common in the southern and western U.S. Most are nocturnal and thus escape notice, even in areas where they are common. In general, these ants are seldom pests, but they occasionally enter homes to look for food. One species, *C. vafer* of southern Arizona, nests in large oak trees. *C. tortuganus* in Florida is also arboreal.

Camponotus (Tanaemyrmex) tortuganus

Camponotus (Tanaemyrmex) tortuganus

3) Subgenus *Myrmentoma*

Like "true" carpenter ants, these ants are usually arboreal, but are also smaller and much less conspicuous. Major and larger media workers can be recognized by the presence of a deep median notch or impression on the anterior border of the clypeus, absent in other *Camponotus*. Most *Myrmentoma* appear adapted for life in dead branches, pine cones, small rotten logs, and hollow plant stems; a couple of species nest in soil. Colonies are small (less than 300 ants) as a rule. These ants are not structural pests, but a few (e.g., *C. nearcticus*) will nest in dwellings and have been known to steal a cookie crumb or two in kitchens.

Camponotus (Myrmentoma) discolor

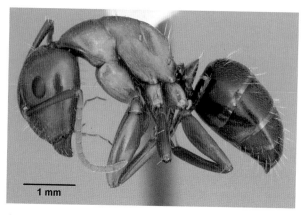

Camponotus (Myrmentoma) discolor

4) Subgenus *Colobopsis*

These tiny species form small arboreal colonies, and their major workers have unique, plug-shaped heads that they use to block nest passages and prevent entry to the nest by non-colony members. They are most common in the southeastern U.S., but a few species are found sporadically in the mountains of the southwest.

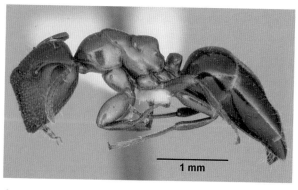

Camponotus (Colobopsis) obliquus

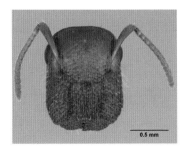

Camponotus (Colobopsis)
obliquus

---

5) Other Subgenera

Members of several other largely tropical species groups also occur in the southern part of our region. The New World *Camponotus* are currently being revised by Bill MacKay. Until the MacKay treatise appears, the best overall references are Snelling (1988) for *Myrmentoma* and Creighton (1950) for the rest of the genus.

CARDIOCONDYLA   Myrmicinae
<10 NA spp.   **INTRODUCED**

DIAGNOSTIC REMARKS: These ants are most easily confused with *Temnothorax.* They may be recognized by the presence of a prominent median clypeal seta, of a distinct metanotal impression, and of a very wide post-petiole, as well as the absence of erect setae on the body.

DISTRIBUTION AND ECOLOGY: All species found in North America are introduced from the Old World. They are small "tramp" ants, commonly transported in commerce, especially in potted plants. Most make tiny, cryptic nests in soil, usually in sunny, disturbed areas. One species is arboreal and nests

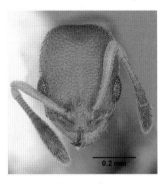

Cardiocondyla emeryi

Cardiocondyla emeryi

in plant cavities. In many species the males occur in two forms, a normal winged morph and a wingless (ergatoid) morph. Colonies are usually small and often highly polygynous. The workers are opportunistic scavengers and predators. Seifert (2003) is the latest revision.

CAREBARA
1 NA sp.

Myrmicinae

DIAGNOSTIC REMARKS: Formerly known as *Oligomyrmex* or *Erebomyrma*, this genus is recognized by the following: the antennae have 11 segments and a 2-segmented apical club, the clypeus is bicarinate, the eyes are minute, and small propodeal teeth are present. The worker caste is dimorphic, and the major workers

Carebara longii

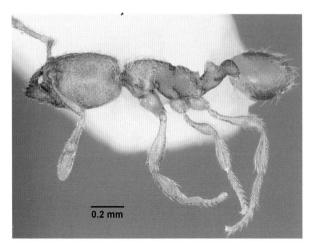

Carebara longii

are enormous compared to the tiny minor workers. The minor workers might be confused with small *Pheidole* minors, but the antennal or clypeal characters above will separate *Carebara* from *Pheidole*. A new revision by Fernandez (2004) contains a key to the New World species.

DISTRIBUTION AND ECOLOGY: Only one rare species *(C. longii)* enters the U.S. in southern and central Texas. It has been collected only two or three times; major workers have not yet been found, and nothing is known about its biology.

CEPHALOTES   Myrmicinae

3 NA spp.

DIAGNOSTIC REMARKS: *Cephalotes* (formerly *Zacryptocerus* for U.S. species) cannot be confused with anything else in the North American ant fauna. The combination of the extremely flattened head; extraordinarily deep antennal scrobes; the large eyes at the posterior corners of the head; the flattened, strongly marginate mesosoma; and the prevalence of coarse foveolate sculpturing on the head and mesosoma is unique among New World ants. A detailed monograph by de Andrade and Baroni Urbani (1999) is the latest source of information on this group.

DISTRIBUTION AND ECOLOGY: Three species of this otherwise neotropical genus just make it into the southern U.S.: one in Florida

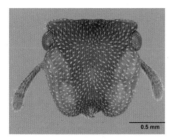

(C. varians), one in Texas *(C. texanus),* and one in southern Arizona *(C. rohweri).* All are arboreal and nest in dead branches or sticks. The peculiar "soldier" subcaste of the workers is adapted to block nest passages with their unique plug-like or cup-shaped heads, which are also present in the queens.

Cephalotes varians (worker)

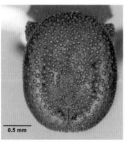

Cephalotes varians (soldier)

CERAPACHYS

2 NA spp.

DIAGNOSTIC REMARKS: Distinctive, heavily sculptured ants, the workers of which are easily recognized by the fully exposed antennal sockets in full-face view, each socket bordered by a prominent lateral carina.

DISTRIBUTION AND ECOLOGY: Only one species has been collected with any frequency. *C. augustae* is known from scattered collections that range from Arkansas to southern California. *C. davisi* is known only from males collected at lights in west Texas and southern New Mexico. *C. augustae*

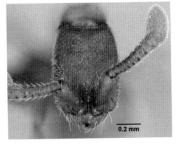

0.2 mm

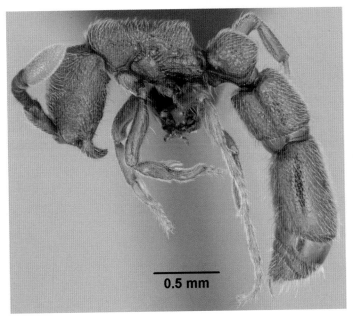

0.5 mm

Cerapachys augustae

may raid other ant nests and feed on their brood. There is also a vague report from California of *C. augustae* workers running in a *Neivamyrmex* column. In the absence of further documentation, this seems most improbable.

CREMATOGASTER   **Myrmicinae**
~30 NA spp.

DIAGNOSTIC REMARKS: An unmistakable genus. The strongly flattened petiole lacking a dorsal node, and the articulation of the

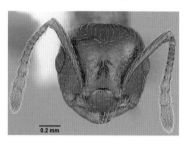

petiole and postpetiole that allows the gaster to point forward over the dorsal surface of the body, set the group apart from other Myrmicinae. Buren (1968) contains the best available keys, but a few problems remain and several undescribed species are not included. An impor-

Crematogaster lineolata

tant new revision of the Costa Rican *Crematogaster* by Longino (2003) makes changes relevant to several North American species.

DISTRIBUTION AND ECOLOGY: Occurs worldwide in temperate and tropical areas. In our area, this genus is most abundant and diverse in the southern states, but two species (*C. lineolata* and *C. cerasi*) have been collected in southern Canada. The majority of our species nest in rotten wood or soil or under rocks, but some species are arboreal. Colonies may be very populous and have multiple queens. Workers are scavengers and predators, tend aphids and coccids on plants, and make extensive use of scent trails. In addition, they are armed with a powerful, repellent defensive secretion that they apply to unfortunate creatures with a flexible, permanently exserted, spatulate sting.

CRYPTOPONE   Ponerinae
2 NA spp.

DIAGNOSTIC REMARKS: Set apart from the superficially similar *Hypoponera* and *Ponera* by the presence of an oval pit on the outer surface of the mandible near the insertion.

DISTRIBUTION AND ECOLOGY: One species, *C. gilva,* is found in soil, litter, and rotten wood in forests of the southeastern and south central U.S. Colonies are small and the ants are probably preda-

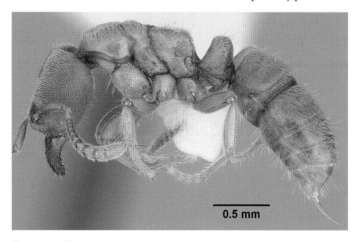

0.5 mm

Cryptopone gilva

Cryptopone gilva

tors of soil microinvertebrates. In 2005, a single ant belonging to a presently unidentified *Cryptopone* species was collected in southeastern Arizona by Alex Wild.

CYPHOMYRMEX   **Myrmicinae**
4 NA spp.

DIAGNOSTIC REMARKS: In this attine genus, the frontal lobes are enlarged laterally so as to form a rounded, flat "plate" on the dorsal surface of the head that covers part of the sides of the head when seen in full-face view. In addition, the body lacks spines and tubercles and is covered by fine-textured, opaque sculpture with numerous short, flattened, appressed hairs.

DISTRIBUTION AND ECOLOGY: Four species are found north of Mexico; perhaps two of these are introductions from the neotropics. All are small, inconspicuous ants. Two species found in the relatively moist habitats of the southeastern U.S. nest in leaf litter and soil; in damp, decaying wood; and under rocks and logs. The two western species live in arid habitats and nest exclusively in soil. They use insect droppings and other animal and plant debris as a substrate for their yeast-like fungus cultures. Best treatments: Kempf (1964) and Snelling and Longino (1992).

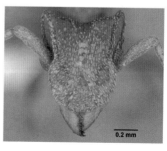

Cyphomyrmex wheeleri

Cyphomyrmex wheeleri

DISCOTHYREA **Proceratiinae**

1 NA sp.

DIAGNOSTIC REMARKS: Like nothing else in the solar system. These minuscule ants have antennal sockets placed on a "shelf" that projects anteriorly over the dorsal surface of small, toothless mandibles. The eyes are vestigial, and the frontal lobes are fused

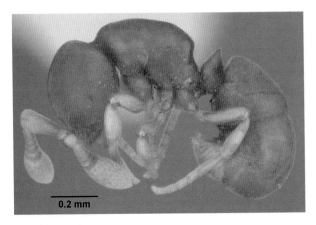

Discothyrea testacea

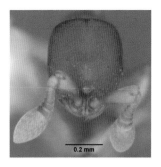

Discothyrea testacea

together to form a small vertical ridge, leaving the antennal sockets completely exposed. The antennae are short but very robust and have a greatly enlarged, football-shaped terminal segment. Lastly, the terminal segments of the gaster project forward under the body, not downward as in almost all other ants. Ants don't get much stranger than this.

DISTRIBUTION AND ECOLOGY: A small genus found throughout the tropics, these ants are seldom seen because of their minute size, subterranean habits, and cryptic coloration. Little is known about their natural history. Our only species, *D. testacea,* is found in the southeastern U.S. and has been collected in leaf litter, humus, and rotten logs. Persistently referred to as "Discos" by the Ant Mafia.

DOLICHODERUS   Dolichoderinae
4 NA spp.

DIAGNOSTIC REMARKS: Recognized by the unique shape of the propodeum. In side view, the dorsal surface forms a horizontal "shelf" that meets the strongly concave posterior surface to form a blunt, posterior-facing ridge. In addition, the petiolar node is relatively thick and blunt in profile. No other North American ants look even vaguely like this except for the exotic *Ochetellus*

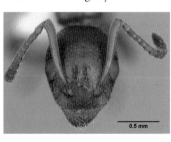

Dolichoderus pustulatus

glaber. In *O. glaber,* however, the posterior face of the propodeum is weakly concave and does not form a horizontal "shelf"; also, the petiolar node is thin and scale-like. MacKay (1993) is the best available reference, but the key in Creighton (1950) still works.

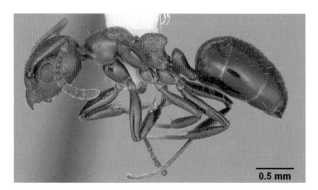

Dolichoderus pustulatus

DISTRIBUTION AND ECOLOGY: These ants are found from eastern Canada south to Florida and west to Minnesota. *D. plagiatus* and *D. pustulatus* form small, cryptic, monogynous colonies in soil, hollow plant stems, and litter, usually in brushy and other semi-open habitats (in the deep south, *D. pustulatus* becomes arboreal). The other two species *(D. taschenbergi, D. mariae)* form large, conspicuous colonies. They make small mounds of vegetative debris, forage along conspicuous trails, and are usually found in bogs or pine barrens. Workers are diurnal foragers and often tend aphids and coccids.

DOLOPOMYRMEX   Myrmicinae
1 NA sp.

DIAGNOSTIC REMARKS: Our only known species, *D. pilatus,* is distinguished by the following combination of characters: median clypeal seta is absent; the antennae are 11-segmented with 3-segmented club; the mandibles have four teeth on a strongly oblique (rather than perpendicular) cutting margin; the

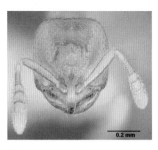

Dolopomyrmex pilatus

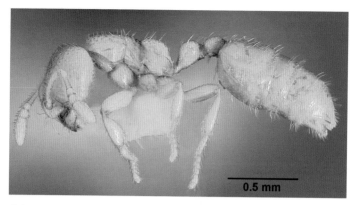

Dolopomyrmex pilatus

eyes are rudimentary or absent; and the propodeum is unarmed. In the field, *D. pilatus* is most likely to be confused with larger *Solenopsis (Diplohoptrum)* species like *S. krockowi*, but the resemblance is purely superficial. This ant belongs to the myrmicine tribe Solenopsidini, even though it lacks a median clypeal seta (Cover and Deyrup 2007).

DISTRIBUTION AND ECOLOGY: These small ants are known from several records that range from southern New Mexico to the Mojave Desert in California. Little is known about their biology, but these ants appear to be almost exclusively subterranean. Sexuals were collected in a nest in southern Arizona in late August. A portion of a colony kept in an artificial nest for a short time fed on young termites as well as ant eggs and brood.

DORYMYRMEX   Dolichoderinae
~20 NA spp.

DIAGNOSTIC REMARKS: In all *Dorymyrmex*, the propodeum has a single, more or less vertical tooth, or "cone," at the juncture of the dorsal and posterior faces. No other North American ants have this propodeal morphology. The eastern forms are well known (Trager 1988), but the western species are poorly understood, and a number remain undescribed. Snelling's (1995) key works for the eastern taxa, but is incomplete with respect to the western

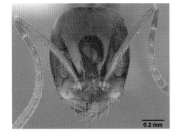

Dorymyrmex insanus

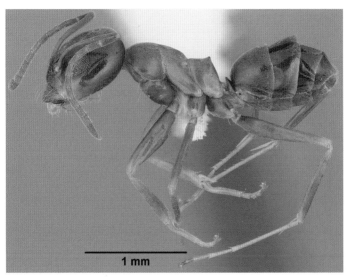

0.2 mm

1 mm

species. Queens are very useful in helping to determine species boundaries in this genus, so collect them whenever you can.

DISTRIBUTION AND ECOLOGY: All *Dorymyrmex* are found in open, xeric habitats, and some are conspicuous elements of open-ground ant faunas throughout the southern and western U.S. These ants are active scavengers, often forage diurnally, and frequently tend aphids and collect fluids from plant nectaries. A few species are temporary social parasites on other congeners, a fascinating phenomenon that has received little study. One species, *D. grandulus*, ranges north to Long Island, New York; another species (perhaps *D. insanus*) has been recorded in North Dakota.

EURHOPALOTHRIX   **Myrmicinae**
1 NA sp.

DIAGNOSTIC REMARKS: Small ants with 7-segmented antennae and deep antennal scrobes that run *below* the eye on the sides of the head. Superficially similar to *Pyramica* species, but easily distinguished by the above characters under proper magnification.

DISTRIBUTION AND ECOLOGY: Most New World species of this genus are tropical. Only one makes it into the U.S.; it occurs in Florida. *E. floridana* is a small, cryptic ant most often collected in the litter and decaying wood of forested habitats. Colonies are small and workers are probably specialized predators.

Eurhopalothrix floridana

FORELIUS   **Dolichoderinae**

~5–6 NA spp.

DIAGNOSTIC REMARKS: In the field, these ants are most likely to be confused with *Tapinoma* or *Linepithema*, but can be readily distinguished by the presence of erect hairs on the pronotal dorsum in *Forelius* (vs. no erect hairs in *Tapinoma* and *Linepithema*). *Forelius* is badly in need of revision in North America. Cuezzo (2000) did not resolve the confusion surrounding these ants in North America, which has been caused by the indiscriminate use of old, poorly defined taxa names.

DISTRIBUTION AND ECOLOGY: Three species occur in the eastern and southeastern United States: *F. pruinosus*; something often called *"analis"* in the older literature; and a very hairy undescribed species known only from Florida. In the west, there are at least two species. *F. mccooki* may be recog-

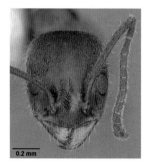

0.2 mm

Forelius n. sp. (Florida)

0.5 mm

Forelius pruinosus

nized by the presence of erect hairs on the antennal scapes. The other western form(s) lack erect hairs on the scapes and may be undescribed. These ants occur only in arid, open habitats. All form large, often polygynous colonies and are especially noteworthy for their ability to forage at extremely high temperatures during the heat of the day. This makes them dominant diurnal foragers in the deserts of the southwestern U.S. and Mexico. Workers make extensive use of trunk trails, and forage on vegetation as well as the soil surface.

FORMICA   Formicinae
>100 NA spp.

DIAGNOSTIC REMARKS: *Formica* are easy to recognize in the field, but the formal characters that separate these ants from the related genus *Lasius* are subtle and can be recognized only with practice (Agosti and Bolton 1990; see also comments under *Lasius*).

DISTRIBUTION AND ECOLOGY: A dominant group in boreal and temperate habitats around the world, this genus is known for its abundance, its ecological importance, and the repeated evolution of social parasitism within the group. We categorize the fauna into species groups.

---

1) *Fusca* Group
Omnipresent, free-living species that range from Labrador and Alaska south into Mexico. Most are black or brownish in color and have at least some distinctive silvery pubescence. Colonies

range from small to large and are often polygynous. These ants are omnivorous and frequently tend aphids and coccids both above and below ground. Francoeur (1973) is essential for identification. This group is sometimes referred to in older literature as the subgenus *Serviformica.*

Formica (fusca group) subsericea

Formica (*fusca* group) *subsericea*

2) *Pallidefulva* Group

An exclusively North American group whose relatively small colonies (usually less than 3,000 ants) are common in open habitats and especially abundant in the eastern and central U.S. The workers are omnivores and forage diurnally. They are hosts for the *Polyergus lucidus* complex of slave-making ants, and are enslaved by some *sanguinea* group *Formica*, most notably *F. pergandei*. A new revision by Trager, MacGown, and Trager (2007) makes reliable identification possible for the first time. This group is often referred to as the subgenus *Neoformica*.

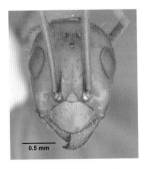

Formica (*pallidefulva* group) *dolosa*

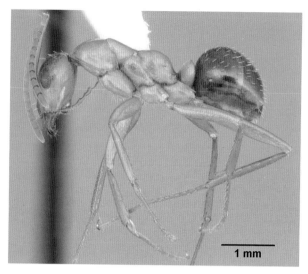

Formica (pallidefulva group) pallidefulva

3) *Neogagates* Group

Ecologically similar to the species of the *pallidefulva* group, these common ants are usually smaller than most other *Formica*; they are also shiny and often brown to black in color. Most species are associated with open habitats, except for the eastern *F. neogagates,* which prefers temperate forest. Colonies are usually small to moderate in size, sometimes polygynous, and the workers are omnivorous. The taxonomy of this group needs work. Some species are quite distinct (e.g., *F. bradleyi, F. limata, F. perpilosa*), but several good species may be lumped under the names "*F. neogagates*" and "*F. lasioides*" at present. Creighton (1950) contains the only available key, and it is still helpful.

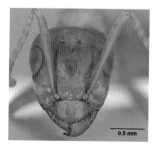

Formica (neogagates group) perpilosa

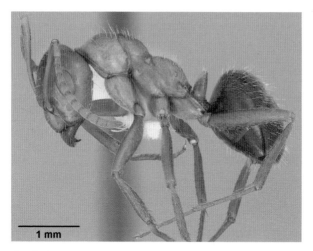

Formica (*neogagates* group) *perpilosa*

4) *Sanguinea* Group

These are facultative slave-making ants characterized by a median concave impression on the anterior border of the clypeus. They enslave other *Formica*, but their workers remain capable of performing all functions necessary to maintain the colony. The queens cannot found colonies alone; they must invade a host nest and secure adoption. *F. sanguinea* group ants are common in a wide variety of habitats and exhibit interesting patterns of host specificity. For example, the eastern *F. rubicunda* and *F. subintegra* enslave only *fusca* group slaves; while *F. creightoni*, true *F. wheeleri*, and *F. gynocrates* enslave only *neogagates* group ants. In contrast, *F. puberula* and *F. pergandei* may enslave *fusca, pallidefulva, neogagates, microgyna,* and *rufa* group *Formica*, and colonies are sometimes found with slaves

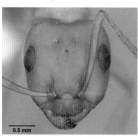

Formica (*sanguinea* group) *gynocrates*

Formica (*sanguinea* group) *gynocrates*

from two or more species groups! Snelling and Buren (1985) is essential for identification. This group is often called *Raptiformica* in the older literature.

5) *Rufa* Group

Known in Europe as "wood ants," members of this group are strongly suspected to be temporary social parasites on a variety of free-living *Formica* species. Single nests containing workers of two *rufa* group species have been collected, suggesting that hyperparasitism (or interspecific colony raiding) may also occur within the group. Colonies are often large, polydomous, aggressive, and polygynous. A few species make conspicuous mounds, but most do not. In any case, nests are often partly or completely thatched with pine needles or other vegetative de-

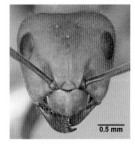

Formica (*rufa* group) *obscuripes*

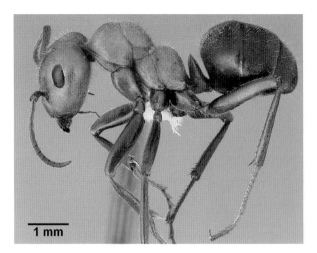

Formica (*rufa* group) *obscuripes*

bris. Workers are omnivorous and forage diurnally. One species, *F. talbotae,* is an inquiline in the nest of the widespread mound-builder *F. obscuripes.* Ants of the *rufa* group are often sporadically distributed, but tend to be a dominant presence where they do occur. The taxonomy of our *rufa* species badly needs updating. Creighton (1950) is still useful.

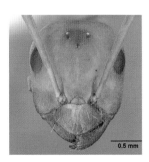

Formica (*microgyna* group) *densiventris*

6) *Microgyna* Group

Exclusively North American, these ants have tiny queens, as small as or even smaller than the largest workers. They are temporary social parasites on other *Formica* species. Otherwise, they are similar ecologically to *rufa* group species, from which they probably evolved. Nests are often made under covering objects, such as grass clumps, wood, stones, or even trash, and thatch

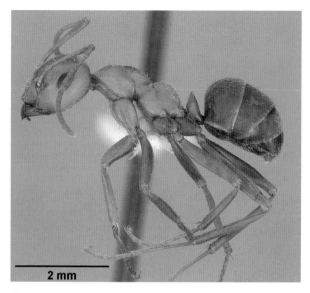

Formica (*microgyna* group) *densiventris*

is commonly present. This is an intriguing group, but there are some taxonomic problems to be solved, plus several new species to be described. Cover is currently working on a revision. Creighton (1950) is still useful.

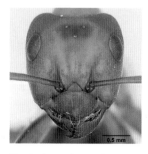

Formica (*exsecta* group) *exsectoides*

7) *Exsecta* Group

Closely related to ants of the *rufa* group, these ants can be recognized by the notably concave posterior border of the head (vs. straight or convex in other *Formica*). Only three species are present in the New World, but they are conspicuous mound-builders and thus attract attention. *F. exsectoides,* the Allegheny Mound-Building Ant, is found

Formica (*exsecta* group) *exsectoides*

sporadically throughout the northern part of the U.S. and southern Canada, west to the front range in Colorado, and south through the Appalachians. *F. ulkei* is a prairie species that occurs from Nova Scotia west through Canada and then south into the Great Plains. *F. opaciventris* is confined to the Rocky Mountain states.

FORMICOXENUS

5 NA spp.

DIAGNOSTIC REMARKS: Closely related to *Leptothorax*, *Formicoxenus* workers and queens can be most easily recognized by the presence of numerous short, erect hairs on the compound eyes (absent in *Leptothorax*). In addition, ergatoid males are present in all species (with alate males also present in some), and intermorphic fe-

Formicoxenus diversipilosus

Formicoxenus diversipilosus

males are common. Francoeur et al. (1985) contains keys to our species and much additional valuable information on these beasts.

DISTRIBUTION AND ECOLOGY: *Formicoxenus* species are unusual "trophic" social parasites. They nest near their hosts, maintaining separate brood chambers. Some species obtain their food in or near the host nest via trophallaxis with host workers. *F. provancheri* and *F. quebecensis* are associated with *Myrmica incompleta* and *M. alaskensis,* respectively. *F. chamberlini* is found with *Manica invidia,* and *F. diversipilosus* and *F. hirticornis* are found in the mound nests of *Formica obscuripes* (and perhaps with other *Formica rufa* group species as well).

GNAMPTOGENYS   **Ectatomminae**

2 NA spp.

DIAGNOSTIC REMARKS: Quite distinct from any other ants in North America. The head, mesosoma, and abdominal segments 3 and 4 (first two gastric segments) are entirely covered by closely parallel costate sculpture, and the body surface is moderately to strongly shiny.

DISTRIBUTION AND ECOLOGY: Two species are present in our region. *G. hartmani* is a small, cryptic, subterranean ant native to Texas and Louisiana. Some evidence suggests that it raids nests of *Trachymyrmex septentrionalis* and eats the brood. *G. triangularis* has been introduced from Central or South America and is now found sporadically in Florida and along the Gulf Coast. It appears to prey upon millipedes. Colonies of both species are small to moderate in size.

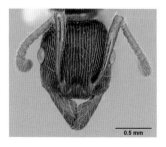

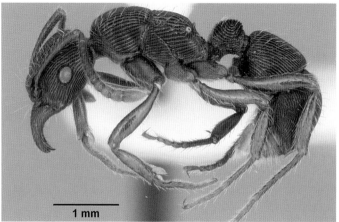

Gnamptogenys triangularis

HARPAGOXENUS

Myrmicinae

1 NA sp.

DIAGNOSTIC REMARKS: Closest in morphology to its *Leptothorax* hosts, *Harpagoxenus* can be readily identified by its enlarged head, prominent antennal scrobes, and mandibles lacking teeth. It can be distinguished from the convergently similar *Protomognathus americanus* by its lack of mandibular teeth (*P. americanus* has four teeth) and other characters.

Harpagoxenus canadensis

Harpagoxenus canadensis

DISTRIBUTION AND ECOLOGY: Our species, *H. canadensis,* is an obligate slave-maker that parasitizes *Leptothorax* species. Its distribution is poorly known. The few records are from cold temperate and boreal areas in central and eastern Canada and the United States, but it may occur in the Rocky Mountains and in western Canada also.

HYPOPONERA   Ponerinae
>7 NA spp.

DIAGNOSTIC REMARKS: Most easily confused with *Ponera*, but it can be distinguished by the simple subpetiolar lobe, which is very different from the more complex lobe found in *Ponera*.

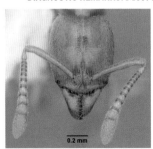

Hypoponera inexorata

DISTRIBUTION AND ECOLOGY: A dominant genus in the litter of tropical forests, *Hypoponera* includes only a few species that range into warm temperate areas. Our species are most abundant in forested habitats of the southeastern United States, but they also

Hypoponera inexorata

occur west to California, mostly in relatively mesic habitats. Colonies are small, and the workers are predators on small soil invertebrates. Some species (e.g., *H. opacior*) produce two types of colonies: one with ordinary winged males and queens, and another with egg-laying workers and wingless, worker-like males that mate with workers emerging from their cocoons. One or two undescribed species are present in the southwestern United States.

LABIDUS

Ecitoninae

1 NA sp.

DIAGNOSTIC REMARKS: Distinguished from *Neivamyrmex* by the presence of a submedian tooth on the tarsal claws (vs. no submedian tooth), and from *Nomamyrmex* by the absence of propodeal teeth.

DISTRIBUTION AND ECOLOGY: Only one species of this neotropical Army ant genus, *L. coecus*, crosses the Mexican border into central and eastern Texas and adjacent states. Colonies are enor-

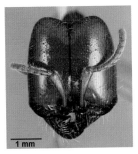

Labidus coecus

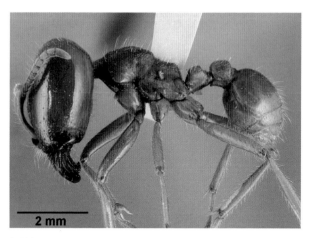

Labidus coecus

mous, but the highly predaceous workers are subterranean foragers and are thus rarely seen above ground.

LASIUS

Formicinae

>56 NA spp.

DIAGNOSTIC REMARKS: *Lasius* look quite different from *Formica* in the field, but the formal characters that separate the two groups are subtle and require practice to recognize. It may be most useful to check the shape of the propodeum in side view. In *Lasius,* the dorsal surface of the propodeum is notably *shorter* than the posterior face. In *Formica,* the propodeum may be evenly convex, or the posterior surface may be equal to or shorter in length than the dorsal surface.

DISTRIBUTION AND ECOLOGY: This genus is virtually omnipresent throughout North America, with the exception of the low, hot deserts of the southwest, as well as peninsular Florida and arctic Canada. All *Lasius* are general scavengers and predators, but also have close relationships with above-ground or subterranean aphids and coccids. Wilson (1955) is an essential reference for understanding our species. This ecologically important genus consists of four species groups.

1) *Niger* Group

These are abundant, diurnal, surface foraging ants whose workers have large compound eyes and are usually brown or yellowish brown in color. In workers, the labial and maxillary palps are long and conspicuous. The queens found colonies independently. This group was previously known as *Lasius,* subgenus *Lasius.*

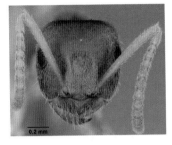

Lasius (*niger* group) *neoniger*

2) *Flavus* Group

This small group consists of subterranean species whose workers have tiny eyes and are yellow in color. These ants tend coccids on plant roots and emerge at the surface only at the time of the mating flights, usually in July through September. The terminal segments of the palps are much reduced, making them short and hard to see. As in the *niger* group, the queens found colonies independently. The *flavus* group is often referred to as the subgenus *Cautolasius.*

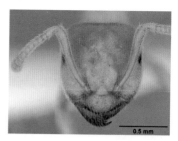

Lasius (flavus group) *flavus*

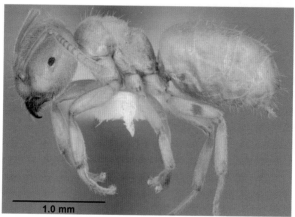

1.0 mm

3) *Umbratus* Group

The queens of these ants are temporary social parasites on *niger* group species and perhaps on other *Lasius* as well. The workers are mostly subterranean but have larger eyes than *flavus* group workers, and are orange, orange brown, or (rarely) yellow in color. Palps are intermediate in length between those of the *niger* and *flavus* groups.

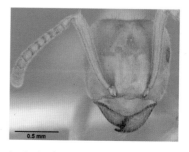

Lasius (umbratus group) *subumbratus*

Lasius (*umbratus* group) *subumbratus*

The common *L. umbratus* is almost certainly a sibling species complex. The *umbratus* group is often referred to as the subgenus *Chthonolasius*.

---

4) *Claviger* Group

Formerly known as the genus *Acanthomyops*, these ants have a reduced palp count of 3,3 instead of 5,3. In the field, they can be

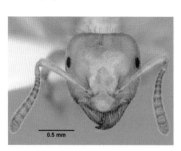

Lasius (*claviger* group) *claviger*

recognized by the following combination of traits: very small eyes, color that ranges from bright yellow to orange, a generally shiny integument surface, and a strong lemony odor (citronella) that is present when the ants are disturbed. Some *umbratus* group species are similar in appearance and

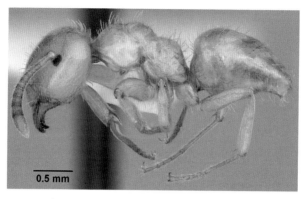

Lasius (claviger group) claviger

may even emit a faint scent of citronella, but the head and meso-soma (at least) are pubescent and not strongly shiny. The workers of *flavus* group species are sometimes confused with *claviger* group workers, but can be distinguished by relatively dense pubescence all over the body, which creates a dull overall appearance. The queens of these ants appear to be temporary social parasites on *Lasius* species, though documentation for this is sparse. As a rule, the workers are subterranean and are seen only when mating flights occur; workers create openings at the soil surface for alates to leave the nest. Workers are omnivorous and typically tend root-feeding aphids and coccids. Mature colonies are large and diffuse, and nest queens are rarely seen. Apparent hybrids occur with some frequency, and have been documented in a useful revision by Wing (1968), which has keys for workers, males, and queens.

LEPTOGENYS   Ponerinae
2 NA spp.

DIAGNOSTIC REMARKS: Unmistakable. The long, toothless, falcate mandibles inserted at the corners of the head are unique among the North American ponerines. The workers also pack a powerful sting, which helps make most encounters with *Leptogenys* even more memorable. See Trager and Johnson (1988) for an account of our species.

DISTRIBUTION AND ECOLOGY: *L. elongata* is known from Louisiana and central and southern Texas. *L. manni* occurs only in northern Florida. Both species form small colonies with a single worker-like (ergatoid) queen. The workers are specialized predators on isopods (pillbugs).

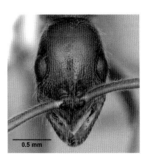

0.5 mm

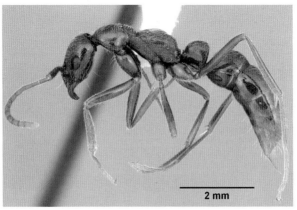

2 mm

Leptogenys manni

LEPTOTHORAX   **Myrmicinae**
10–15 NA spp.

DIAGNOSTIC REMARKS: These ants have mandibles with six teeth, 11 antennal segments, at least a small metanotal impression, and a petiole that lacks a distinct anterior peduncle. At the species level, the taxonomy is a mess, in large part because most of the species are weakly differentiated morphologically. Some new species await description, but no real progress will be possible until the presently recognized forms are better defined. Older names for the genus are *Leptothorax* subgenus *Mycothorax* and *muscorum* species group.

DISTRIBUTION AND ECOLOGY: This boreal to cool temperate group occurs in relatively cold habitats. Found coast to coast in northern North America, *Leptothorax* species extend south at high altitudes to southern Arizona and southern California, and probably into northern Mexico. Often common in appropriate habitats,

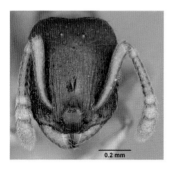

they nest in preformed cavities in rotten wood, under stones, in soil, and in hollow plant stems. Colonies are often polygynous and polydomous. Some species in this group are hosts to social parasites, including the slave-making *Harpagoxenus canadensis,* the inquilines *Leptothorax paraxenus, L. wilsoni, L. faberi,* and *L. pocahontas,* and perhaps others.

0.2 mm

Leptothorax retractus

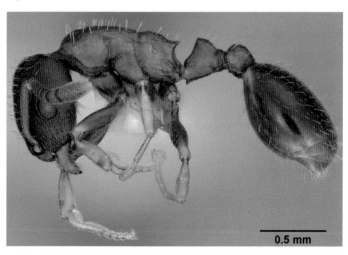

0.5 mm

Leptothorax n. sp. (Arizona)

LINEPITHEMA

Dolichoderinae

1 NA sp.

DIAGNOSTIC REMARKS: Readily distinguished from the workers of other dolichoderines in North America by the propodeum, which is evenly convex in profile and separated from the promesonotum by a distinct metanotal impression. Erect hairs are absent on the mesosomal dorsum, and the petiolar scale is well developed.

DISTRIBUTION AND ECOLOGY: This neotropical genus is represented in the U.S. by the introduced pest *L. humile*, the Argentine Ant. This miserable beast is found in coastal southern and central California, and sporadically along the Gulf Coast. It is an important pest, especially in urban situations, invading homes in search of food and

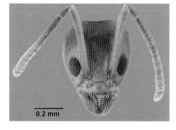

0.2 mm

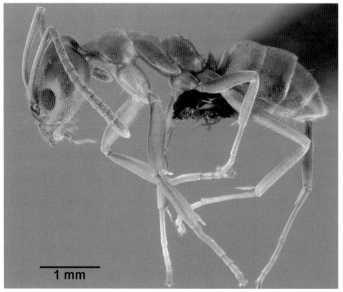

1 mm

Linepithema humile

water and tending various garden aphids and coccids for their secretions. In many situations, it exterminates native ant species where it becomes established. Another species, *L. iniquum,* has been recorded from greenhouses several times, but has not yet become established outdoors in our area.

LIOMETOPUM   **Dolichoderinae**
3 NA spp.

DIAGNOSTIC REMARKS: Recognized by the following combination of characters: workers are slightly to moderately polymorphic; the mesosomal profile is continuous and evenly convex; and erect hairs are abundant on the mesosomal dorsum. The workers also emit a pungent smell (reminiscent of blue cheese) when disturbed.

DISTRIBUTION AND ECOLOGY: Three species occur in the western and southwestern U.S. They are associated with oak forests, oak-juniper woodlands, riparian forests and moderate-elevation, pine-dominated forests. All form enormous colonies in hollow trees or in the soil and forage along conspicuous trails. The workers are active, abundant, aggressive, and armed with disagreeable defensive compounds. Creighton's (1950) key to our species still works.

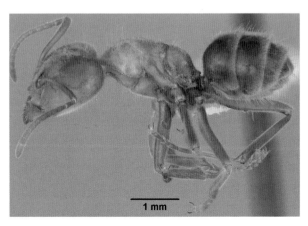

Liometopum apiculatum

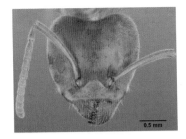

Liometopum apiculatum

0.5 mm

MANICA

Myrmicinae

4 NA spp.

DIAGNOSTIC REMARKS: Easily recognized by the following combination of characters: the palp formula is 6,4; the mandibles have 12 or more teeth or denticles; the propodeum is unarmed; and the petiolar node rounded in profile. In the field, these ants look like unusually large *Myrmica* (which have six to 10 mandibular teeth and propodeal spines or teeth).

DISTRIBUTION AND ECOLOGY: Four species of this small, Holarctic genus occur in western North America. Two species (*M. invidia* and *M. hunteri*) occur in the Rocky Mountain region; two others

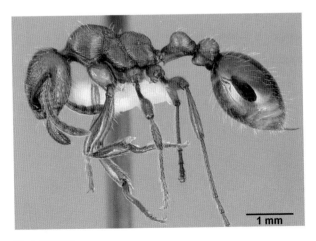

1 mm

Manica hunteri

occur in California's Sierra Nevada region *(M. bradleyi* and *M. parasitica)*. *M. invidia* and *M. hunteri* are found in open grassland or sagebrush habitats and riparian forest edges. *M. bradleyi* lives in coniferous forests and forest edges. *M. parasitica* is a little-known social parasite on *M. bradleyi*. The free-living species

Manica hunteri

are all generalized predators and scavengers. See Wheeler and Wheeler (1970) for an account of our species.

MESSOR
9 NA spp.

Myrmicinae

DIAGNOSTIC REMARKS: The formal character separating *Messor* and the closely related genus *Aphaenogaster* is the large metasternal process, which is absent in *Aphaenogaster*. In the field, however, *Messor* is more likely to be confused with *Pogonomyrmex*. The easiest way to separate them is by the profile of the mesosoma. In *Messor*, the propodeum is always distinctly depressed below the level of the promesonotum, and a pronounced metanotal impression is often present. In *Pogonomyrmex*, the mesoso-

Messor chicoensis

Messor chicoensis

mal profile is pretty much evenly convex, and the propodeum is never notably depressed below the level of the promesonotum. Referred to as *Veromessor* in much of the older literature. M. R. Smith (1956) has a key to species.

DISTRIBUTION AND ECOLOGY: Most of our *Messor* occur in California (Norte and Sur) and adjacent states of the U.S. and Mexico. *M. lobognathus* is an exception and occurs sporadically east to North Dakota. *Messor* inhabit a wide variety of arid and semi-arid habitats, and seeds are a large component of their diets. The worker caste is continuously polymorphic in some species, and almost monomorphic in others. Colonies vary in size with species. For example, *M. pergandei,* an abundant species in hot desert regions of southern Arizona, California, and northern Mexico, forms enormous colonies with large, conspicuous craters. *M. smithi,* from southern California and Nevada, forms much smaller colonies with modest nests that are easy to overlook. Johnson (2000, 2001) are two important general references.

MONOMORIUM   Myrmicinae
~16 NA spp.

DIAGNOSTIC REMARKS: Abundant and diverse in the Old World, *Monomorium* is represented by only a small number of species in North America, some of which are exotic. *Monomorium* workers are most often confused with those of *Solenopsis.* Both have a bicarinate clypeus and a median clypeal seta. Luckily, *Monomorium* workers have 11-segmented antennae with a 3-segmented antennal club whereas *Sole-*

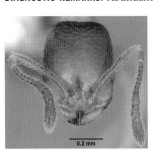

Monomorium destructor

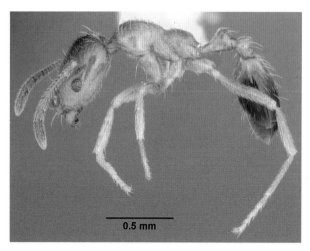

Monomorium destructor

nopsis have 10-segmented antennae with a 2-segmented club. The native species all belong to the *minimum* species group. Workers of the *minimum* group can be readily recognized in the field. They are tiny (less than 2 mm long), slender, jet black, and shiny. DuBois (1986) is unsatisfactory, but provides the only recent treatment of the *minimum* group. Cover is currently working on a new revision.

DISTRIBUTION AND ECOLOGY: Colonies of all free-living species are often polygynous and polydomous. The ants nest in rotten wood, under rocks, under bark, and in soil. *M. pergandei* and *M. talbotae* are fabulously rare inquilines in nests of the widespread *M. minimum.* Some *minimum* group species produce only winged queens, while others produce only apterous queens. In at least one other species *(M. emersoni)*, both apterous and alate queens occur. Other species may also produce both winged and apterous queens—a phenomenon that cries out for investigation. *M. pharaonis* (the Pharaoh Ant) is a pantropical tramp from Africa that can be an enormous nuisance in heated buildings. *M. floricola* is another pantropical tramp that is common in central and southern Florida and occasionally infests buildings.

MYCETOSORITIS   **Myrmicinae**

1 NA sp.

DIAGNOSTIC REMARKS: This small attine genus can be confused with *Cyphomyrmex* or *Trachymyrmex*. Unlike *Cyphomyrmex*, *Mycetosoritis* lacks expanded frontal lobes covering the sides of the head in front of the eyes (in frontal view), and it *has* erect hairs on the body. Separating it from *Trachymyrmex* is a bit more problematic. North American *Trachymyrmex* species have tubercles on the mesosoma and the gaster. Neither condition occurs in *Mycetosoritis hartmanni*.

DISTRIBUTION AND ECOLOGY: Only one species *(M. hartmanni)* occurs in the U.S., from western Louisiana to southern Texas.

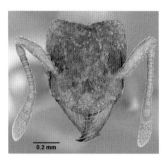

Colonies are small (usually less than 100 ants), and the cryptic nests are built in sandy soil. A population observed by one of the authors (SPC) was in very sandy soil, in a small, sunny gap in mature oak-pine forest. The nests were surmounted by tiny turrets. The best treatment is Wheeler (1907).

Mycetosoritis hartmanni

MYRMECINA Myrmicinae

2 NA spp.

DIAGNOSTIC REMARKS: Distinguished from other North American myrmicines by the following characters: the petiole is nearly cylindrical, with a node that is absent or rudimentary, and a prominent carina runs the length of the head, at or close to the ventral margin of the side.

DISTRIBUTION AND ECOLOGY: Primarily an Asian group, two (perhaps three) species of *Myrmecina* occur in the U.S. and Canada. *M. americana* ranges from southern Canada to Baja California. *M. californica,* presently a synonym of *M. americana,* occurs in central and northern California and may also be a good species.

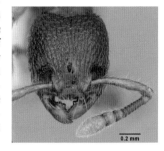

In addition, a rare, undescribed social parasite occurs in nests of *M. americana* throughout its range. All of our species are found mostly in wooded areas, nesting in small colonies in soil and litter. These ants appear to be predators of soil microinvertebrates and oribatid mites in particular.

Myrmecina americana

MYRMECOCYSTUS   **Formicinae**

~30 NA spp.

DIAGNOSTIC REMARKS: Distinguished from other formicines by greatly elongate maxillary palps in which segment 4 is longer than segments 5 and 6 taken together. In addition, the workers of most species have a prominent psammophore on the ventral surface of the head.

DISTRIBUTION AND ECOLOGY: This is a strictly North American genus known from the western United States and Mexico. These ants, together with those of the genus *Pogonomyrmex,* are dominant elements in desert ant faunas. While the myrmicine genus relies on seeds as a storable food resource, many *Myrmecocystus* species derive much of their sustenance from plant exudates, mainly sap and nectar from floral and extrafloral glands, and

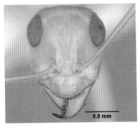

from aphid and coccid secretions. This liquid is stored within the greatly distended gasters of large workers, called "repletes." Foragers are also active predators and scavengers, with termites being a favored food. Snelling's texts (1976, 1982) are required for identification of species.

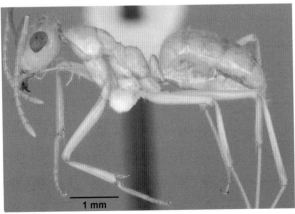

Myrmecocystus navajo

MYRMELACHISTA Formicinae
1 NA sp.

`INTRODUCED`

DIAGNOSTIC REMARKS: *M. ramulorum* has a 9-segmented antennae with a 3-segmented apical club. *Brachymyrmex* also has 9-segmented antennae, but the apical club is absent.

DISTRIBUTION AND ECOLOGY: *M. ramulorum* was introduced into Florida from the Caribbean region, but doubts remain as to

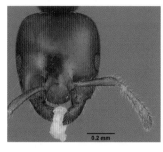

whether it has established itself in the U.S. Other species of this exclusively arboreal genus are found in Central and South America. Habits are poorly known, but colonies are often small and the ants probably feed on plant-derived carbohydrates supplemented by dead invertebrates, usually scavenged.

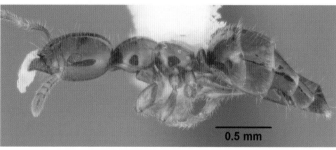

Myrmelachista ramulorum

MYRMICA Myrmicinae
50–75 NA spp.

DIAGNOSTIC REMARKS: The following characters are important in recognizing this genus: the palp count is 6,4; the propodeum is armed with teeth or spines; the petiole lacks a distinct peduncle; and the tibial spurs are finely pectinate. In most species, the head and mesosoma are covered by heavy rugose sculpture. The taxonomic history of the group in North America is troubled, to say the least! Just how many species there are is anyone's guess. Bolton (1994) lists about 25 taxa, but the number of good species should be double that. Francoeur (in prep.) is working on a much-needed new revision.

DISTRIBUTION AND ECOLOGY: *Myrmica* is ecologically important and abundant in boreal communities and many temperate habitats. Colonies are usually small to moderate in size, often polygnous, and nest in soil, under rocks, or in rotten wood. The ants are primarily carnivorous, but also tend aphids and take plant sap

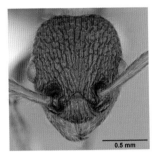

and flower nectar. Three extremely rare species are inquilines in nests of other *Myrmica*. The Eurasian *M. rubra* has become established in scattered localities in New England, and it has recently been collected in Seattle, Washington. The ant is a pest because of its aggressive biting and stinging, which is apparently not a problem in its native Europe.

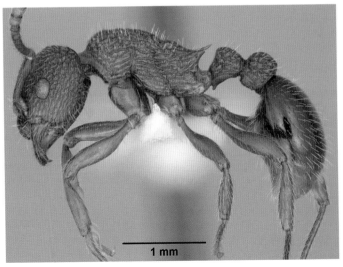

Myrmica n. sp. (Arizona)

NEIVAMYRMEX Ecitoninae

~25 NA spp.

DIAGNOSTIC REMARKS: Distinguished from *Labidus* and *Nomamyrmex* by its simple tarsal claws (vs. tarsal claws with a submedian tooth). In the field, it can be an unforgettable sight, with its long, fast-running columns of foraging or emigrating ants, many carrying brood or prey. For years, *Neivamyrmex* taxonomy has been cursed with a parallel system of names caused by the propensity of early taxonomists to name males without knowledge of the associated workers. Snelling and Snelling (2007) cleans up some of the redundancy caused by this practice and provides new keys and updated information about these fascinating beasts.

DISTRIBUTION AND ECOLOGY: The only truly temperate army ants, *Neivamyrmex* are found across much of the central and southern U.S., peaking in diversity and abundance in the desert southwest. The nomadic and predaceous life-history of these ants is well known, and the sizable colonies host a large number of myrmecophiles. Most *Neivamyrmex* are predominantly subterranean but will forage on the soil surface at night, or on cloudy days after rain. Some species target the brood of other ant species. Males frequently come to lights at night.

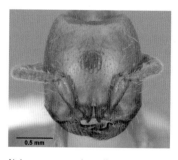
Neivamyrmex swainsonii

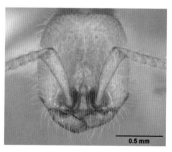
Neivamyrmex harrisii

Neivamyrmex swainsonii

Neivamyrmex harrisii

NESOMYRMEX   **Myrmicinae**

1 NA sp.

DIAGNOSTIC REMARKS: *Nesomyrmex wilda* is the only species of this small pantropical genus found north of Mexico. It closely resembles a yellow *Temnothorax*, but can be easily distinguished by the presence of tubercles on the petiole (absent in *Temnothorax*).

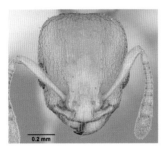

Kempf (1959) presents a key to *Nesomyrmex* species.

DISTRIBUTION AND ECOLOGY: *N. wilda* is arboreal and nests in dead sticks in trees and shrubs. It occurs in the Rio Grande valley near Brownsville, Texas, and south into Mexico. Colonies are relatively small.

Nesomyrmex wilda

NOMAMYRMEX   **Ecitoninae**

1 NA sp.

DIAGNOSTIC REMARKS: In the context of the North American ant fauna, *Nomamyrmex* are the only army ants with both propodeal teeth and tarsal claws with a submedian tooth. *Nomamyrmex* also have short scapes and a robust, heavily armored build. Though seldom seen, their foraging columns resemble a Panzer division on the march. See Watkins (1985) for more information.

DISTRIBUTION AND ECOLOGY: One species *(N. esenbeckii,* sometimes referred to as *N. esenbeckii wilsoni)* barely gets north of the Mexican border into central and southern Texas. These highly subterranean ants are strongly suspected of being specialized predators of *Atta*—about as tough a method for putting dinner on the table as any found among the Animalia! *Nomamyrmex* ants have also been seen raiding *Camponotus* nests.

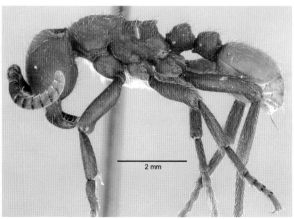

Nomamyrmex esenbeckii

OCHETELLUS   **Dolichoderinae**

1 NA sp.   **INTRODUCED**

DIAGNOSTIC REMARKS: Distinguished by its unusual propodeal structure: the posterior face is weakly concave in profile, and the juncture with the dorsal face is angulate. It might possibly be confused with *Dolichoderus,* but see comments under that genus.

DISTRIBUTION AND ECOLOGY: One arboreal species, *O. glaber,* of this Australasian genus has been introduced into parts of north-central Florida.

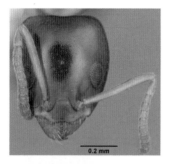

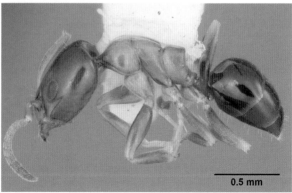

Ochetellus glaber

ODONTOMACHUS   **Ponerinae**

5 NA spp.

DIAGNOSTIC REMARKS: The unique head shape and mandibles inserted just on either side of the cephalic midline set these ants apart from nearly all other ponerines. In the U.S. the single tooth or spine at the apex of the petiole separates *Odontomachus* from the closely related genus *Anochetus* (two teeth). For the North American species, see Deyrup and Cover (2004a). Brown (1976) is essential for a world view of the genus.

DISTRIBUTION AND ECOLOGY: Our *Odontomachus* species range from Florida and Georgia west to southern Arizona. Colonies may be found in mesic or dry habitats, are usually small (usually less than 200 ants), and normally have a single queen. The workers are effective predators, employing highly specialized "trap-jaw" mandibles to kill or disable their prey.

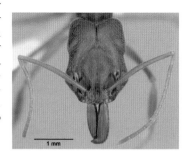

Odontomachus relictus

PACHYCONDYLA   Ponerinae

4 NA spp.

DIAGNOSTIC REMARKS: This large, heterogeneous, pantropical genus is in a state of taxonomic confusion, but the North American species are well resolved except for *P. stigma,* which may be a complex of species. *Pachycondyla* are most easily confused with *Hypoponera,* but *Pachycondyla* have two tibial spurs on the middle and hind legs (vs. one in *Hypoponera*).

DISTRIBUTION AND ECOLOGY: The large *P. villosa* occurs only in extreme southern Texas, nests in hollow trees and logs, and is an active predator with a nasty sting. *P. harpax* occurs in Texas and Louisiana, nests primarily in soil, and is seldom collected. *P.*

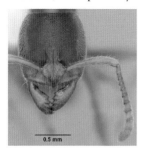

stigma occurs in Florida, nesting in rotten wood. This New World ant has become established in parts of the South Pacific and Asian tropics. The Asian *P. chinensis* occurs sporadically in the middle Atlantic states and is now forming dense populations in parts of South Carolina (E. Paysen, unpublished collection data).

Pachycondyla stigma

PARATRECHINA   **Formicinae**

~20 NA spp.

DIAGNOSTIC REMARKS: Small formicine ants with five (rarely six) mandibular teeth. They're most easily recognized by the presence of long, coarse setae, often black or brown, arranged in conspicuous pairs on the mesosomal dorsum.

DISTRIBUTION AND ECOLOGY: This cosmopolitan genus is represented in North America by about 20 species; five of them are exotics. All nest in soil, but some will also nest in litter, in rotting wood, and in dwellings. A few species (e.g., *P. longicornis, P. bourbonica*) can be household pests. *Paratrechina* nest in a wide range of habitats, and several occur in deserts or desert grasslands. Three undescribed inquiline social parasites are now known from the eastern U.S. (Cover, unpublished collection data). Trager (1984) is the best modern reference.

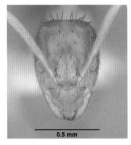

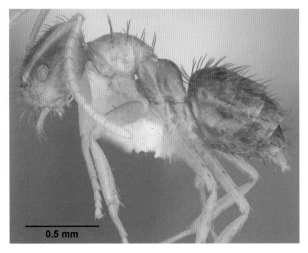

Paratrechina terricola

PHEIDOLE   **Myrmicinae**

100 NA spp.?

DIAGNOSTIC REMARKS: Worker caste dimorphic (strongly poly-morphic in a small minority of species). The following characters are helpful in recognizing these ants: antennae are 12-segmented with a distinct 3-segmented club (rarely 4-segmented: *P. clydei, P. grundmanni*); the propodeum is always depressed below the level of the promesonotum; the palp formula is 2,2 or 3,2; and the propodeum usually has teeth or spines. Wilson's majestic tome (2003) is the very latest reference. Gregg (1959) remains useful for the North American species.

DISTRIBUTION AND ECOLOGY: Only a small fraction of the enor-mous neotropical *Pheidole* fauna enters the U.S. Wilson (2003) delimits 624 species in the New World but lists just 73 from the U.S. The true number of species present is probably close to 100, as Cover and Johnson (unpublished collection data) already have 15 additional undescribed species from Arizona alone, and new ones continue to be discovered nearly every year. Some species are notable seed harvesters, but the majority are predaceous or omnivorous. Two, *P. inquilina* and *P. elecebra,* are inquilines in the nests of *P. coloradensis* and *P. ceres,* respectively. Pheidole is a dominant ant genus in many habitats across the southern U.S.

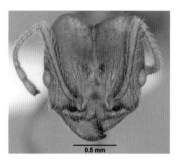
Pheidole hyatti (major worker)

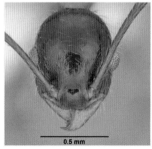
Pheidole hyatti (minor worker)

Pheidole hyatti (major worker)

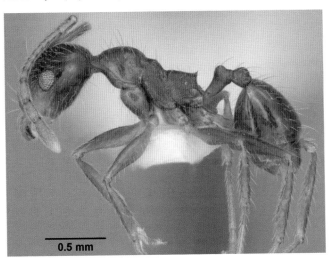

Pheidole hyatti (minor worker)

PLAGIOLEPIS
Formicinae
1 NA sp.

DIAGNOSTIC REMARKS: In our area, tiny yellow formicines with 11-segmented antennae and no erect hairs on the mesosomal dorsum.

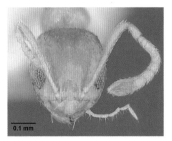

DISTRIBUTION AND ECOLOGY: This is an Old World genus; most of the species appear to be Afrotropical. *P. alluaudi,* a widespread tropical tramp ant, has been recorded from Santa Catalina Island in California. We do not know if it has become established there.

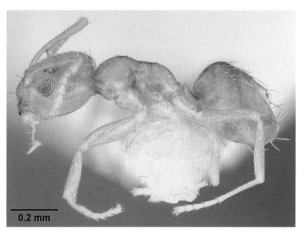

Plagiolepis alluaudi

PLATYTHYREA   **Ponerinae**

1 NA sp.

DIAGNOSTIC REMARKS: Best recognized in the field by its elongate body, peculiar matte gray appearance, fast running speed, and notable sting. *Platythyrea* might be confused with a *Pachycondyla,* but the antennal sockets are widely separated and the tibial spurs are comb-like (vs. antennal sockets set close together and tibial spurs simple in *Pachycondyla*).

DISTRIBUTION AND ECOLOGY: Our only species, *P. punctata,* has a circum-Caribbean distribution and enters our area only in southern Florida and extreme southern Texas. Colonies are small and inconspicuous, but the ant has a fascinating life-history (see Schilder, Heinze, and Holldobler [1999] and references therein for the scoop).

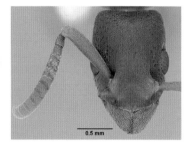

Platythyrea punctata

POGONOMYRMEX   **Myrmicinae**

25 NA spp.

DIAGNOSTIC REMARKS: Can be recognized by the following characters: the middle and hind tibial spurs are finely pectinate (comb-like); the petiole has a long anterior peduncle; the node is sharply conical with a short, nearly vertical anterior face; in most species, the head and mesosoma are covered by fine parallel costate sculpture; and psammophore is usually present. Cole's (1968) monograph is *the* masterpiece, but Taber (1998) has updated keys that include some recently described species.

DISTRIBUTION AND ECOLOGY: These are the well-known harvester ants of the American West, whose colonies are such a conspicuous part of arid landscapes in the U.S. and Mexico. The colonies of some species are moderate to large in size, and the workers can be aggressive foragers for seeds and insects. Other species make much smaller nests and have relatively docile workers. *Note:* The workers of the larger species are renowned for their potent stings. Most "Pogos" are found in arid and semi-arid regions of the central and western U.S., but one species *(P. badius)* occurs in the southeastern states; it is also our only polymorphic species. Three species belonging to the old subgenus *Ephebomyrmex* occur in the desert southwest. These differ radically from their congeners, as they are small ants that fashion cryptic nests and exhibit highly modified life-histories. Johnson 2000 and 2001 are important general references.

Pogonomyrmex badius

Pogonomyrmex imberbiculus

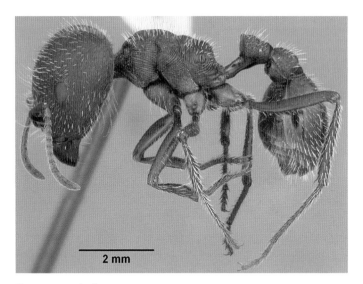

Pogonomyrmex badius

Pogonomyrmex imberbiculus

POLYERGUS
Formicinae

2–7 NA spp.

DIAGNOSTIC REMARKS: The sharp, sickle-shaped, toothless mandibles are unique among the formicines. The workers are generally blood red in color, but occasionally the gaster is darker. The number of species in North America is uncertain, but it could be as high as six or seven; a modern revision is badly needed.

DISTRIBUTION AND ECOLOGY: These are obligate slave-making ants that prey upon species of the related genus *Formica*. Our *Polyergus* sort into two groups: *lucidus* complex ants, which use *pallidefulva* group *Formica* as hosts, and *rufescens* complex species,

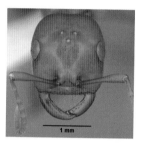

which use *fusca* group *Formica* as slaves. *Polyergus* are not common as a rule, but can be locally abundant where they do occur. They seem to require large, relatively stable host populations and are apparently vulnerable to local extinction where host populations are subject to frequent or severe disturbance.

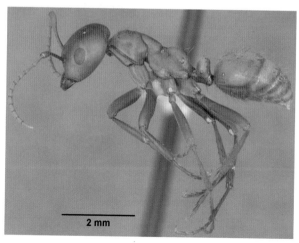

Polyergus breviceps

PONERA   **Ponerinae**

2 NA spp.

DIAGNOSTIC REMARKS: Small, relatively inconspicuous ants that can be distinguished from other from other ponerines by their unique subpetiolar process, as described in the key.

DISTRIBUTION AND ECOLOGY: Both North American *Ponera* species are inhabitants of soil, litter, and rotten wood, primarily in forested habitats of the eastern United States. *P. pennsylvanica* ranges from eastern and southern Canada to the Florida panhandle and west to the limits of the eastern deciduous forest. The unfortunately named *P. exotica* is native, not exotic, and it is known from scattered collections in the southeastern U.S. Colonies are small, and the workers forage in soil and litter for a variety of small soil invertebrates. *P. pennsylvanica* is abundant in many forest communities in the eastern U.S.

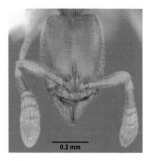

0.2 mm

0.5 mm

Ponera exotica

PRENOLEPIS
Formicinae

1 NA sp.

DIAGNOSTIC REMARKS: Shiny, reddish brown to dark brown ants reminiscent of large *Paratrechina*, but they can be recognized by the fine erect setae on the mesosomal dorsum (vs. paired, coarse, bristle-like setae on the mesosomal dorsum in *Paratrechina*). They might also be mistaken for *neogagates* group *Formica*, but they have six mandibular teeth (instead of seven or more teeth or denticles in *Formica*).

DISTRIBUTION AND ECOLOGY: Only one species, *P. imparis*, is presently recognized from North America. It is quite variable and has an enormous geographic range that stretches from southern

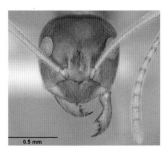

Canada to California and south into central Mexico. Colonies forage during the spring and fall and estivate during the hottest months. During these dormant periods, the ants subsist on fats stored in "replete" workers. *Prenolepis* workers are renowned for their ability to forage at cold temperatures

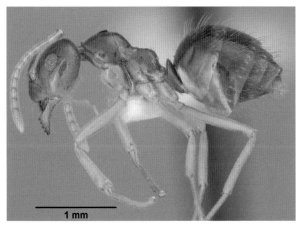

Prenolepis imparis

not tolerated by other ants. Despite this, the distribution of *P. imparis* is decidedly temperate, rather than boreal. An analysis of geographic variation in these ants is much needed; *P. imparis* may be a sibling species cluster.

PRIONOPELTA
1 NA sp.

Amblyoponinae

INTRODUCED

DIAGNOSTIC REMARKS: As in *Amblyopone,* the anterior clypeal margin is denticulate, but in *Prionopelta* the mandibles are short and have three teeth (vs. long, falcate, and with many teeth in *Amblyopone*).

DISTRIBUTION AND ECOLOGY: One species of this small, pantropical genus, *P. antillana,* has been introduced into central Florida, where it has been collected in litter samples.

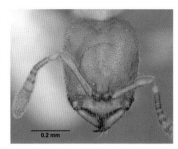

Prionopelta antillana

The neotropical species of this genus are all rather uniform in morphology and ecology. Nests are in rotten wood or soil, and colonies contain up to 700 ants. The small workers are subterranean foragers and prey on diplurans and other small soil invertebrates (Hölldobler and Wilson 1986).

PROCERATIUM Proceratiinae

8 NA spp.

DIAGNOSTIC REMARKS: An unusual genus recognized by the following combination of characters: the tergite of abdominal segment 4 (second gastric tergite) is enlarged so as to make the remaining 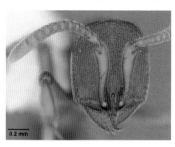 abdominal segments project forward under the body (rather than downward or rearward as in almost all other ants); the frontal lobes are small and closely approximated (leaving the antennal sockets partly exposed when seen from above);

Proceratium chickasaw

and the mandibles have several teeth (vs. no teeth, as in *Discothyrea*). Baroni Urbani and de Andrade (2003) is an unparalleled source of information about these ants.

DISTRIBUTION AND ECOLOGY: *Proceratium* are inconspicuous nesters in soil or rotten wood. Most species prefer mesic, forested habitats. Colonies are very small (less than 50 ants), and the workers are specialist predators of the eggs of other arthropods. *Proceratium* is most common and diverse in the southeastern U.S., but occurs in Texas and California as well. Specimens are most often collected by litter sifting.

PROTOMOGNATHUS Myrmicinae
1 NA sp.

DIAGNOSTIC REMARKS: Recognized by the combination of mandibles with four teeth, a median concavity present on the anterior clypeal border, and the presence of well-developed antennal scrobes. This ant is most easily confused with the convergently sim-

Protomognathus americanus

ilar *Harpagoxenus canadensis,* but *Harpagoxenus* has toothless mandibles.

DISTRIBUTION AND ECOLOGY: One species, *P. americanus,* is known from the eastern U.S. and southeastern Canada, where it is an obligate slave-making ant attacking three closely related species of *Temnothorax (T. ambiguus, T. curvispinosus,* and *T. longispinosus).* The host species are most common in forested habitats, but also occur in overgrown old fields, where they nest in dead stems of *Apocynum* and *Asclepias* and in old goldenrod galls.

PSEUDOMYRMEX   Pseudomyrmecinae
10 NA spp.

DIAGNOSTIC REMARKS: Unmistakable. The combination of enormous compound eyes, very short antennae, arboreal nests, and wasp-like habitus and activity make these ants readily recognizable in the field. Ward (1985) resolves many old taxonomic problems and presents a key to the species in North America.

DISTRIBUTION AND ECOLOGY: *Pseudomyrmex* species are found coast to coast in southern parts of the United States. These ants are arboreal, nesting in dead or living branches or twigs in trees or shrubs, but some species will also nest in grass culms and in herbaceous plant stems. Generally, colonies are small or moderate in size (less than 500 ants) and often occupy several nest sites.

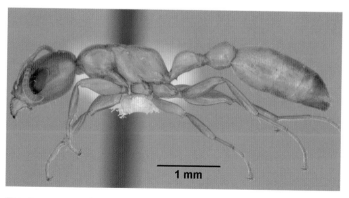

Pseudomyrmex apache

The workers are omnivorous and frequently forage diurnally. *P. apache* and *P. pallidus* are adapted to life in arid climates, and can be found in fairly xeric habitats in Arizona and California. *P. leptosus* is an inquiline social parasite in nests of *P. ejectus*. It has been collected only in Florida.

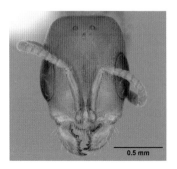

Pseudomyrmex apache

PYRAMICA   **Myrmicinae**
40 NA spp.

DIAGNOSTIC REMARKS: Pyramica is a heterogeneous group of species that share the following characters: 1) antennae are 6-segmented; 2) the mandibles usually have a finely dentate cutting margin occupying part or all of the visible length of the mandible; 3) setae are often scale-like or spatulate; and 4) the petiole and postpetiole usually have prominent, light-colored, sponge-like processes. Bolton's (2000) world revision contains keys to the North American species.

DISTRIBUTION AND ECOLOGY: This genus, as redefined by Bolton (1999, 2000), includes the species formerly placed in the genera *Epitritus, Trichoscapa,* and *Smithistruma*. Nearly 40 species occur in the United States, mostly in the forests of the southeastern states, where they nest in soil, leaf litter, or in rotten wood. Most appear to be specialized predators of collembolans, but other small arthropods may also be taken. A few exotic forms are now established in the southeastern U.S. A handful of rare species occur in Arizona and California. One, *P. arizonica*, is associated with nests of *Trachy-*

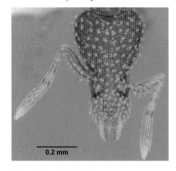

Pyramica creightoni

Pyramica creightoni

myrmex arizonensis, where it preys upon Collembola that abound there and benefits from the climate control the *Trachymyrmex* provide for their fungus gardens.

ROGERIA **Myrmicinae**
3 NA spp.

DIAGNOSTIC REMARKS: Small ants with the following characters: the clypeus is bicarinate; the mesosomal dorsum, seen in profile, is evenly convex; the metanotal impression is absent; the antennae have a distinct, 3-segmented club; and the petiolar spiracle is located on the peduncle, not on the anterior face of the node. Kugler's (1994) world revision has a key to species.

DISTRIBUTION AND ECOLOGY: Three species are known from the southern U.S. (Texas, Arizona, and California); the one from California is undescribed. Virtually nothing is known about the biology of these ants, which have been collected just a handful of times in the U.S.

Rogeria creightoni

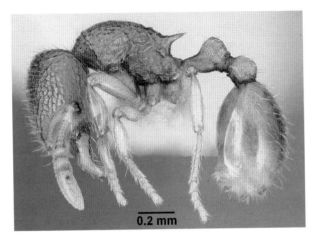

Rogeria creightoni

SOLENOPSIS Myrmicinae

~40 NA spp.

DIAGNOSTIC REMARKS: Easily recognized by the 10-segmented antennae with a 2-segmented apical club. Other important characters: the clypeus is bicarinate; the carinae usually end as teeth on the anterior clypeal border; a median clypeal seta is present; and the propodeum is unarmed.

DISTRIBUTION AND ECOLOGY: Our species fall into three groups.

1) *Geminata* Group

Commonly known as Fire Ants, these species form enormous colonies, have a polymorphic worker caste, and sting like the dickens. There are four native species in the southern U.S. (*S. amblychila*, *S. aurea*, *S. geminata*, and *S. xyloni*) and two imported species,

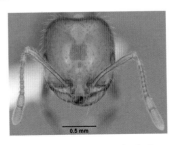

Solenopsis (*geminata* group) *xyloni*

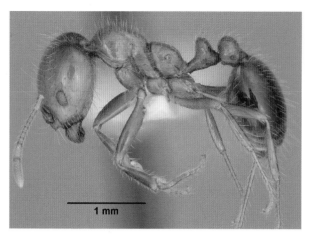

Solenopsis (*geminata* group) *xyloni*

S. richteri and the notorious Red Imported Fire Ant, *S. invicta.* Trager (1991) has keys to the species.

2) Subgenus *Diplorhoptrum*

Known as "thief ants," these ants have tiny workers and often live in close proximity to colonies of larger ants, preying on their brood and carting off their stored food. A few "Diplos" are arboreal. There are many species in North America (some undescribed) and their taxonomy is very poorly understood. Thompson's (1989a, 1989b) short papers are essential for Florida. For the rest of the continent, in the absence of anything better, see Creighton (1950).

3) Subgenus *Euophthalma*

In North America, this group is represented on the Gulf Coast by the circum-Caribbean *S. globularia,* a small ant easily distinguished from other congeners by its extraordinarily broad postpetiole.

STENAMMA   **Myrmicinae**

>17 NA spp.

DIAGNOSTIC REMARKS: Can be recognized by the following characters: the clypeus is bicarinate; the carinae don't end as teeth on the anterior clypeal margin; the petiole has a peduncle; the spiracle is located on the peduncle, not on the anterior face of the node; a metanotal impression is present; the propodeum has short teeth; and the antennal club is 4-segmented or indistinct. M.R. Smith (1957) treats the eastern *Stenamma;* Snelling (1973) treats the western species.

DISTRIBUTION AND ECOLOGY: Inconspicuous soil-nesting ants usu-

ally found in forests and woodlands, *Stenamma* occur throughout the U.S. and Canada and south into Central America. Colonies are usually small (less than 200 ants) and monogynous. *Stenamma* are almost certainly predators of soil microinvertebrates. They are most commonly collected by litter sifting.

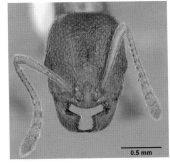

Stenamma fovolocephalum

STRUMIGENYS   **Myrmicinae**

5 NA spp.

DIAGNOSTIC REMARKS: Similar to *Pyramica,* except that the mandibles lack a finely dentate cutting edge and instead have a terminal fork of two or three inward-facing teeth. Sometimes one or two pairs of small subapical teeth are present as well, but the mandibles never appear to have a cutting edge. These mandibular characters are simply the most visible evidence of the profound differences in the structure of the mouthparts that differentiate *Pyramica* from *Strumigenys.* Bolton's (2000) mono-

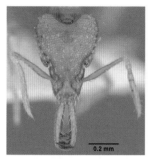

graph is an essential resource.

DISTRIBUTION AND ECOLOGY: *S. louisianae* is our only native species. The other species are all tramps, most of them established in peninsular Florida. Colonies are small and are usually found in leaf litter, soil, or rotten wood. These ants are specialist predators on Collembola.

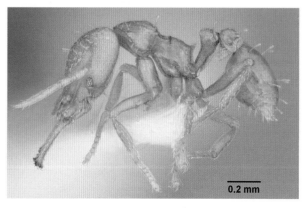

Strumigenys rogeri

TAPINOMA

Dolichoderinae

>5 NA spp.

DIAGNOSTIC REMARKS: Easily recognized by the unique morphology of the gaster and by the absence of a petiolar scale. In *Tapinoma,* there are only four gastric tergites (representing abdominal segments 4 to 6). The fifth tergite (abdominal segment 7) is reflexed ventrally, thus locating the anal pore on the ventral surface of the gaster, not at the terminus as in other dolichoderines.

DISTRIBUTION AND ECOLOGY: Three free-living species occur in the United States. *T. sessile* is native and has perhaps the widest geographic range and greatest ecological tolerance of any ant in North America. It occurs from coast to coast, from southern Canada south to Florida, south into the Mexican highlands, and in almost every conceivable habitat type, including buildings, upon occasion. It is also quite variable in size

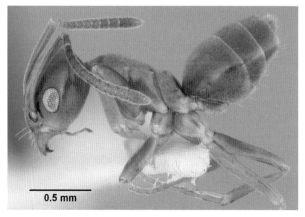

Tapinoma sessile

and color. Genetic analysis may reveal it to be a sibling species complex. *T. sessile* is sometimes a minor pest in homes. *T. melanocephalum* has been introduced into Florida, probably from the Asian tropics. Known as the "Ghost Ant," it infests buildings with enthusiasm, and it will also nest outside if the climate is subtropical. The third species, *T. litorale,* nests in dead sticks on trees in southern Florida and the Caribbean. New evidence strongly suggests that *T. dimmocki,* presently listed as a synonym of *T. sessile,* is in fact an inquiline parasite of that species. In addition, another undescribed inquiline has been found in nests of *T. sessile* in Colorado (Cover, unpublished collection data).

TECHNOMYRMEX   Dolichoderinae
1 NA sp.   **INTRODUCED**

DIAGNOSTIC REMARKS: An Old World genus, often superficially similar to *Tapinoma.* However, it lacks the distinctive abdominal morphology as described for that genus, and thus the anal pore is located at the gastric terminus.

DISTRIBUTION AND ECOLOGY: A single species has been introduced into southern Florida and southern California. Until the present, it has been referred to as *T. albipes,* the notorious "White-footed Ant" that has spread throughout much of the tropics and subtropics in recent years. Recent work by Barry Bolton has revealed

Technomyrmex difficilis

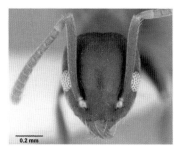

that the ant present in the United States is a closely related species, whose proper name is *T. difficilis.* Pending completion of the study, the U.S. form should be referred to as "*Technomyrmex* cf. *albipes.*" This beast is a nuisance in residential areas. It nests in houses and abounds in gardens and landscaped areas, where it tends aphids and coccids on an assortment of cultivated plants.

Technomyrmex difficilis

TEMNOTHORAX   **Myrmicinae**
~50 NA spp.

DIAGNOSTIC REMARKS: Until very recently, these ants were lumped into *Leptothorax* as the subgenus *Myrafant.* These ants have five mandibular teeth and either 11 or 12 antennal segments. Only one of our species, the peculiar *T. pergandei,* has a pronounced

Temnothorax smithi

metanotal impression; all other species lack any trace of an impression at the metanotum. MacKay (2000) is the latest monograph.

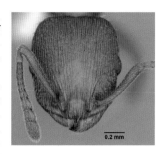

Temnothorax smithi

DISTRIBUTION AND ECOLOGY: Widespread throughout the region, these ants nest in a wide variety of nest sites: in soil, litter, preformed cavities (such as acorns and hickory nuts), rotten wood, and hollow plant stems and sticks, or under stones and the bark of living or dead trees. Colonies are generally small, consisting of less than 100 to 200 ants. Some species are monogynous, but others are polygynous and sometimes polydomous. The workers forage individually and are "shy" (they avoid competition with other ants). They are opportunistic scavengers and predators of soil microinvertebrates, and occasionally tend aphids and collect plant sap. A few species occur in arid grassland or desert habitats and collect seeds. *T. duloticus* is a slave-making ant that parasitizes three species common in eastern North America, *T. ambiguus, T. curvispinosus,* and *T. longispinosus.*

TETRAMORIUM   **Myrmicinae**
10 NA spp.

DIAGNOSTIC REMARKS: Best recognized by the peculiar structure of the antennal sockets. In *Tetramorium,* the posterior border of the clypeus drops nearly vertically to form a deep crater or pit entirely surrounding the antennal insertion. In other superficially similar genera (e.g., *Myrmica, Leptothorax,* and *Pogonomyrmex* [*Ephebomyrmex*]), the antennal socket does not take the form of a deep pit. The posterior border of the clypeus slopes gradually towards the

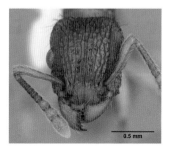

Tetramorium bicarinatum

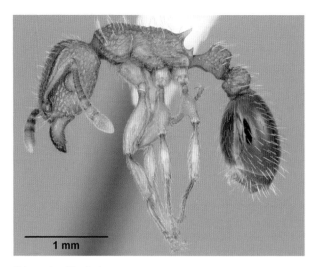

Tetramorium bicarinatum

antennal insertion, creating a shallow depression anteriorly. Bolton (1979) is required for identification.

DISTRIBUTION AND ECOLOGY: This cosmopolitan genus is represented in the region by two species native to the deserts of the southwestern U.S. and northern Mexico, as well as seven introduced forms. Four of the introduced species are tropical; two are temperate. *T. caespitum* occurs widely throughout temperate North America. *T. tsushimae,* a close relative, has been reported recently from around St. Louis, Missouri. All are omnivores, but are otherwise variable in their natural history.

TRACHYMYRMEX   Myrmicinae

7–10 NA species

DIAGNOSTIC REMARKS: Uncomfortably similar to the closely related attine genus *Acromyrmex*. *Trachymyrmex* workers are generally monomorphic, as opposed to polymorphic in *Acromyrmex*. In addition, *Acromyrmex* workers of all sizes have true spines on the promesonotum. *Trachymyrmex* workers generally have low tubercles that are often covered with strongly curved setae. In some species, the mesosomal tubercles can seem spine-like, but they are still strongly tuberculate and have coarse, recurved setae; both characters are absent in *Acromyrmex versicolor*. Lastly, in some *Trachymyrmex,* the frontal carinae extend back nearly to the rear corners of the head, creating a very shallow antennal scrobe; this is never seen in *A. versicolor.* The taxonomy of the western species needs revision.

DISTRIBUTION AND ECOLOGY: *Trachymyrmex* is mostly a neotropical group, but one species is found in the eastern United States. *T. septentrionalis* occurs north to the New Jersey and Long Island Pine Barrens along the Atlantic coast and west to Texas. Another six (or more) occur in the southwestern states, from central Texas to Arizona. As is true of other attines, these ants cultivate fungi on

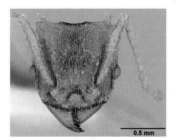

a vegetable compost within chambers in their subterranean nests. Temperate species seem especially fond of oak catkins as a substrate for their fungus cultures. Colonies are small to moderate in size, and nests are sometimes cryptic.

Trachymyrmex septentrionalis

VOLLENHOVIA

1 NA sp.

DIAGNOSTIC REMARKS: The petiole lacks an anterior peduncle and has a large, ventral translucent plate. No other myrmicine in our area has such a structure.

DISTRIBUTION AND ECOLOGY: One species of this East Asian genus, *V. emeryi*, has become established in and around the District of Columbia, with a single record from Philadelphia. *V. emeryi* is native to Japan and Korea. In Washington, D.C., it occurs in riparian forest and at the bottoms of moist creek valleys in very moist, rotten wood. Colonies are polygynous and polydomous, and the queens are brachypterous. *V. emeryi* was probably imported with the 3,020 cherry trees given to the people of the

United States as a gift by the Japanese government in 1912 and planted around the Tidal Basin. *Paratrechina flavipes*, another Asian exotic common in D.C., may have arrived in the same shipment.

Vollenhovia emeryi

WASMANNIA

Myrmicinae

1 NA sp.

INTRODUCED

DIAGNOSTIC REMARKS: A remarkable ant recognized by the following characters: the frontal carinae are prominent, extending nearly to the rear border of the head and forming shallow antennal scrobes; the petiole has a distinct peduncle, with node sharply quadrate in profile and dorsal surface separated from the anterior and posterior faces by distinct angles; the antennae have 11 segments and a 2-segmented club; and the eyes have a flat ventral side.

DISTRIBUTION AND ECOLOGY: This poorly understood neotropical genus consists of perhaps a dozen species. The best-known species is the so-called Little Fire Ant, *W. auropunctata,* that has become widely spread by commerce and is now a major pest in

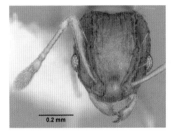

several areas of the tropics (e.g., the Galapagos, New Caledonia, parts of west Africa). In the U.S., the ant is established in southern Florida. Colonies are polygynous, unicolonial, and nest in soil, litter, and dead wood, under bark and

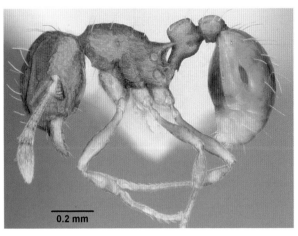

Wasmannia auropunctata

trash, around houses ... in other words, almost everywhere. The tiny orange workers are omnivorous and forage using scent trails. Note: The workers are reputed to pack a sting all out of proportion to their small size, but many myrmecologists have collected them repeatedly without ever having been stung.

XENOMYRMEX   **Myrmicinae**
1 NA sp.

DIAGNOSTIC REMARKS: Slender, shiny arboreal ants bearing a superficial resemblance to a *Monomorium* or *Solenopsis (Diplorhoptrum)* species, but readily distinguished by the subcylindrical petiole and rudimentary dorsal node. Our only species is *X. floridanus,* which also occurs in parts of the Caribbean.

DISTRIBUTION AND ECOLOGY: This small neotropical genus enters our area only in central and southern Florida. The ants nest in dead branches in tree canopies. Colonies appear to be small to moderate in size and may be polydomous.

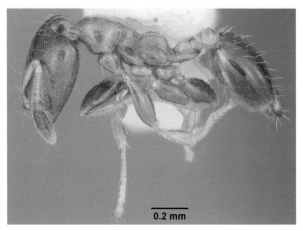

Xenomyrmex floridanus

ANT GENERA OF NORTH AMERICA BY SUBFAMILY

Note: Asterisks (*) denote introduced genera.

AMBLYOPONINAE

Amblyopone
*Prionopelta**

CERAPACHYINAE

Acanthostichus
Cerapachys

DOLICHODERINAE

Dolichoderus
Dorymyrmex
Forelius
*Linepithema**
Liometopum
*Ochetellus**
Tapinoma
*Technomyrmex**

ECITONINAE

Labidus
Neivamyrmex
Nomamyrmex

ECTATOMMINAE

Gnamptogenys

FORMICINAE

Acropyga
Brachymyrmex

Camponotus
Formica
Lasius
Myrmecocystus
*Myrmelachista**
Paratrechina
*Plagiolepis**
Polyergus
Prenolepis

MYRMICINAE

Acromyrmex
*Anergates**
Aphaenogaster
Atta
*Cardiocondyla**
Carebara
Cephalotes
Crematogaster
Cyphomyrmex
Dolopomyrmex
Eurhopalothrix
Formicoxenus
Harpagoxenus
Leptothorax
Manica
Messor
Monomorium
Mycetosoritis
Myrmecina

Myrmica
Nesomyrmex
Pheidole
Pogonomyrmex
Protomognathus
Pyramica
Rogeria
Solenopsis
Stenamma
Strumigenys
Temnothorax
Tetramorium
Trachymyrmex
Vollenhovia*
Wasmannia*
Xenomyrmex

PONERINAE

Anochetus*
Cryptopone
Hypoponera
Leptogenys
Odontomachus
Pachycondyla
Platythyrea
Ponera

PROCERATIINAE

Discothyrea
Proceratium

PSEUDOMYRMECINAE

Pseudomyrmex

ANT SPECIES
NORTH OF MEXICO:
A WORKING LIST

This informal working list is based primarily on Bolton (1995), supplemented by the relevant literature up to the present and, in a few instances, unpublished results (or results in press) from ongoing research, both that of SPC and others. Undescribed species, of which there are no small number, are not listed here. This is a working list, not a formal checklist, thus it is not to be made the basis for changes in taxonomic status. Genera and species are listed alphabetically.

Acanthostichus arizonensis MacKay
Acanthostichus punctiscapus MacKay
Acanthostichus texanus Forel
Acromyrmex versicolor (Pergande)
Acropyga epedana Snelling
Amblyopone oregonensis (Wheeler)
Amblyopone orizabana Brown
Amblyopone pallipes (Haldeman)
Amblyopone trigonignatha Brown
Anergates atratulus (Schenck)
Anochetus mayri Emery
Aphaenogaster albisetosa Mayr
Aphaenogaster ashmeadi (Emery)
Aphaenogaster boulderensis
    M. R. Smith
Aphaenogaster carolinensis Wheeler
Aphaenogaster cockerelli Andre
Aphaenogaster flemingi M. R. Smith
Aphaenogaster floridana M. R. Smith
Aphaenogaster fulva Roger

Aphaenogaster huachucana Creighton
Aphaenogaster lamellidens Mayr
Aphaenogaster mariae Forel
Aphaenogaster megommata
    M. R. Smith
Aphaenogaster miamiana Wheeler
Aphaenogaster occidentalis (Emery)
Aphaenogaster patruelis Forel
Aphaenogaster picea (Wheeler)
Aphaenogaster puncticeps MacKay
Aphaenogaster rudis Enzmann
Aphaenogaster tennesseensis (Mayr)
Aphaenogaster texana Wheeler
Aphaenogaster treatae Forel
Aphaenogaster uinta Wheeler
Aphaenogaster umphreyi Deyrup
    & Davis
Aphaenogaster valida Wheeler
Atta mexicana (F. Smith)
Atta texana (Buckley)

Brachymyrmex depilis Emery
Brachymyrmex obscurior Forel
Brachymyrmex patagonicus Mayr
Camponotus absquatulator Snelling
Camponotus acutirostris Wheeler
Camponotus americanus Mayr
Camponotus anthrax Wheeler
Camponotus bakeri Wheeler
Camponotus caryae (Fitch)
Camponotus castaneus (Latreille)
Camponotus chromaiodes Bolton
Camponotus clarithorax Creighton
Camponotus cuauhtemoc Snelling
Camponotus decipiens Emery
Camponotus discolor (Buckley)
Camponotus dumetorum Wheeler
Camponotus essigi M. R. Smith
Camponotus etiolatus Wheeler
Camponotus festinatus (Buckley)
Camponotus floridanus (Buckley)
Camponotus fragilis Pergande
Camponotus herculeanus (Linnaeus)
Camponotus hyatti Emery
Camponotus impressus (Roger)
Camponotus laevigatus (F. Smith)
Camponotus mccooki Forel
Camponotus maritimus Ward
Camponotus microps Snelling
Camponotus mina Forel
Camponotus mississippiensis
    M. R. Smith
Camponotus modoc Wheeler
Camponotus nearcticus Emery
Camponotus novaeboracensis (Fitch)
Camponotus obliquus M. R. Smith
Camponotus ocreatus Emery
Camponotus papago Creighton
Camponotus pennsylvanicus (DeGeer)
Camponotus planatus Roger
Camponotus pudorosus Emery
Camponotus pylartes Wheeler
Camponotus quercicola M. R. Smith
Camponotus sansabeanus (Buckley)
Camponotus sayi Emery
Camponotus schaefferi Wheeler
Camponotus semitestaceus Snelling
Camponotus sexguttatus (Fabricius)
Camponotus snellingi Bolton

Camponotus socius Roger
Camponotus subbarbatus Emery
Camponotus texanus Wheeler
Camponotus tortuganus Emery
Camponotus trepidulus Creighton
Camponotus ulcerosus Wheeler
Camponotus vafer Wheeler
Camponotus vicinus Mayr
Camponotus yogi Wheeler
Cardiocondyla emeryi Forel
Cardiocondyla mauritanica Forel
Cardiocondyla minutior Forel
Cardiocondyla obscurior Wheeler
Cardiocondyla venustula Wheeler
Cardiocondyla wroughtonii (Forel)
Carebara longii (Wheeler)
Cephalotes rohweri (Wheeler)
Cephalotes texanus (Santschi)
Cephalotes varians (F. Smith)
Cerapachys augustae Wheeler
Cerapachys davisi M. R. Smith
Crematogaster ashmeadi Mayr
Crematogaster atkinsoni Wheeler
Crematogaster browni Buren
Crematogaster californica Wheeler
Crematogaster cerasi (Fitch)
Crematogaster coarctata Mayr
Crematogaster colei Buren
Crematogaster crinosa Mayr
Crematogaster dentinodis Forel
Crematogaster depilis Wheeler
Crematogaster distans Mayr
Crematogaster emeryana Creighton
Crematogaster hespera Buren
Crematogaster isolata Buren
Crematogaster laeviuscula Mayr
Crematogaster larreae Buren
Crematogaster lineolata (Say)
Crematogaster marioni Buren
Crematogaster minutissima Mayr
Crematogaster missuriensis Emery
Crematogaster mormonum Wheeler
Crematogaster mutans Buren
Crematogaster navajoa Buren
Crematogaster nocturna Buren
Crematogaster obscurata Emery
Crematogaster opaca Mayr
Crematogaster opuntiae Buren

Crematogaster pilosa Emery
Crematogaster pinicola Deyrup
    & Cover
Crematogaster punctulata Emery
Crematogaster rifelna Buren
Crematogaster smithi Creighton
Crematogaster torosa Mayr
Crematogaster vermiculata Emery
Cryptopone gilva (Roger)
Cyphomyrmex flavidus Pergande
Cyphomyrmex minutus Mayr
Cyphomyrmex rimosus (Spinola)
Cyphomyrmex wheeleri Forel
Discothyrea testacea Roger
Dolichoderus mariae Forel
Dolichoderus plagiatus (Mayr)
Dolichoderus pustulatus Mayr
Dolichoderus taschenbergi Mayr
Dolopomyrmex pilatus Cover
    & Deyrup
Dorymyrmex bicolor Wheeler
Dorymyrmex bossutus (Trager)
Dorymyrmex bureni (Trager)
Dorymyrmex elegans (Trager)
Dorymyrmex flavopectus M. R. Smith
Dorymyrmex flavus McCook
Dorymyrmex grandulus (Forel)
Dorymyrmex insanus (Buckley)
Dorymyrmex lipan Snelling
Dorymyrmex medeis Trager
Dorymyrmex paiute Snelling
Dorymyrmex reginiculus (Trager)
Dorymyrmex smithi Cole
Dorymyrmex wheeleri (Kusnezov)
Eurhopalothrix floridana Brown
    & Kempf
Forelius mccooki (McCook)
Forelius pruinosus (Roger)
Formica accreta Francoeur
Formica adamsi Wheeler
Formica aerata Francoeur
Formica alpina Wheeler
Formica altipetens Wheeler
Formica archboldi M. R. Smith
Formica argentea Wheeler
Formica aserva Forel
Formica biophilica Trager, MacGown
    & Trager

Formica bradleyi Wheeler
Formica calviceps Cole
Formica canadensis Santschi
Formica ciliata Mayr
Formica coloradensis Creighton
Formica comata Wheeler
Formica creightoni Buren
Formica criniventris Wheeler
Formica curiosa Creighton
Formica dakotensis Emery
Formica densiventris Viereck
Formica difficilis Emery
Formica dirksi Wing
Formica dolosa Buren
Formica emeryi Wheeler
Formica exsectoides Forel
Formica ferocula Wheeler
Formica foreliana Wheeler
Formica fossaceps Buren
Formica francoeuri Bolton
Formica fuliginothorax Blacker
Formica fusca Linnaeus
Formica glacialis Wheeler
Formica gnava Buckley
Formica gynocrates Snelling & Buren
Formica hewitti Wheeler
Formica impexa Wheeler
Formica incerta Buren
Formica indianensis Cole
Formica integra Nylander
Formica integroides Wheeler
Formica knighti Buren
Formica laeviceps Creighton
Formica lasioides Emery
Formica lepida Wheeler
Formica limata Wheeler
Formica longipilosa Francoeur
Formica lugubris Zetterstedt
Formica manni Wheeler
Formica microgyna Wheeler
Formica microphthalma Francoeur
Formica moki Wheeler
Formica montana Wheeler
Formica morsei Wheeler
Formica mucescens Wheeler
Formica neoclara Emery
Formica neogagates Viereck
Formica neorufibarbis Emery

Formica nepticula Wheeler
Formica nevadensis Wheeler
Formica obscuripes Forel
Formica obscuriventris Mayr
Formica obtusopilosa Emery
Formica occulta Francoeur
Formica opaciventris Emery
Formica oreas Wheeler
Formica oregonensis Cole
Formica pacifica Francoeur
Formica pallidefulva Latreille
Formica pergandei Emery
Formica perpilosa Wheeler
Formica planipilis Creighton
Formica podzolica Francoeur
Formica postoculata Kennedy
    & Dennis
Formica prociliata Kennedy
    & Dennis
Formica propinqua Creighton
Formica puberula Emery
Formica querquetulana Kennedy
    & Dennis
Formica ravida Creighton
Formica reflexa Buren
Formica rubicunda Emery
Formica scitula Wheeler
Formica sibylla Wheeler
Formica spatulata Buren
Formica subelongata Francoeur
Formica subintegra Wheeler
Formica subnitens Creighton
Formica subpolita Mayr
Formica subsericea Say
Formica talbotae Wilson
Formica transmontanis Francoeur
Formica ulkei Emery
Formica vinculans Wheeler
Formica wheeleri Creighton
Formica whymperi Wheeler
Formica xerophila M. R. Smith
Formicoxenus chamberlini (Wheeler)
Formicoxenus diversipilosus
    (M. R. Smith)
Formicoxenus hirticornis (Emery)
Formicoxenus provancheri Emery
Formicoxenus quebecensis Francoeur

Gnamptogenys hartmani (Wheeler)
Gnamptogenys triangularis (Mayr)
Harpagoxenus canadensis M. R. Smith
Hypoponera gleadowi (Forel)
Hypoponera inexorata (Wheeler)
Hypoponera opaciceps (Mayr)
Hypoponera opacior (Forel)
Hypoponera punctatissima (Roger)
Labidus coecus (Latreille)
Lasius alienus (Foerster)
Lasius arizonicus Wheeler
Lasius atopus Cole
Lasius bureni (Wing)
Lasius californicus Wheeler
Lasius claviger (Roger)
Lasius colei (Wing)
Lasius coloradensis Wheeler
Lasius creightoni (Wing)
Lasius crypticus Wilson
Lasius fallax Wilson
Lasius flavus (Fabricius)
Lasius humilis Wheeler
Lasius interjectus Mayr
Lasius latipes (Walsh)
Lasius minutus Emery
Lasius murphyi (Forel)
Lasius nearcticus Wheeler
Lasius neoniger Emery
Lasius nevadensis Cole
Lasius niger (Linnaeus)
Lasius occidentalis Wheeler
Lasius pallitarsis (Provancher)
Lasius plumopilosus Buren
Lasius pogonogynus Buren
Lasius pubescens Buren
Lasius sitiens Wilson
Lasius speculiventris Emery
Lasius subglaber Emery
Lasius subumbratus Viereck
Lasius umbratus (Nylander)
Lasius vestitus Wheeler
Lasius xerophilus MacKay & MacKay
Leptogenys elongata (Buckley)
Leptogenys manni Wheeler
Leptothorax acervorum (Fabricius)
Leptothorax calderoni Creighton
Leptothorax crassipilis Wheeler

Leptothorax faberi Buschinger
Leptothorax muscorum (Nylander)
Leptothorax paraxenus Heinze &
    Alloway
Leptothorax pocahontas (Buschinger)
Leptothorax retractus Francoeur
Leptothorax sphagnicola Francoeur
Leptothorax wilsoni Henize
Linepithema humile (Mayr)
Liometopum apiculatum Mayr
Liometopum luctuosum Wheeler
Liometopum occidentale Emery
Manica bradleyi (Wheeler)
Manica hunteri (Wheeler)
Manica invidia Bolton
Manica parasitica (Creighton)
Messor andrei (Mayr)
Messor chamberlini Wheeler
Messor chicoensis (M. R. Smith)
Messor lariversi (M. R. Smith)
Messor lobognathus Andrews
Messor pergandei (Mayr)
Messor smithi (Cole)
Messor stoddardi (Emery)
Monomorium cyaneum Wheeler
Monomorium destructor (Jerdon)
Monomorium ebeninum Forel
Monomorium emarginatum DuBois
Monomorium emersoni Gregg
Monomorium ergatogyna Wheeler
Monomorium floricola (Jerdon)
Monomorium minimum (Buckley)
Monomorium pergandei (Emery)
Monomorium pharaonis (Linnaeus)
Monomorium talbotae DuBois
Monomorium trageri DuBois
Monomorium viride Brown
Mycetosoritis hartmanni Wheeler
Myrmecina americana Emery
Myrmecocystus christineae Snelling
Myrmecocystus colei Snelling
Myrmecocystus creightoni Snelling
Myrmecocystus depilis Forel
Myrmecocystus ewarti Snelling
Myrmecocystus flaviceps Wheeler
Myrmecocystus hammettensis Cole
Myrmecocystus kathjuli Snelling

Myrmecocystus kennedyi Snelling
Myrmecocystus koso Snelling
Myrmecocystus lugubris Wheeler
Myrmecocystus melliger Forel
Myrmecocystus mendax Wheeler
Myrmecocystus mexicanus Wesmael
Myrmecocystus mimicus Wheeler
Myrmecocystus navajo Wheeler
Myrmecocystus placodops Forel
Myrmecocystus pyramicus M. R. Smith
Myrmecocystus romainei Snelling
Myrmecocystus semirufus Emery
Myrmecocystus snellingi Bolton
Myrmecocystus tenuinodis Snelling
Myrmecocystus testaceus Emery
Myrmecocystus wheeleri Snelling
Myrmecocystus yuma Wheeler
Myrmelachista ramulorum Wheeler
Myrmica alaskensis Wheeler
Myrmica americana Weber
Myrmica brevispinosa Wheeler
Myrmica colax (Cole)
Myrmica crassirugis Francoeur
Myrmica detritinodis Emery
Myrmica discontinua Weber
Myrmica emeryana Forel
Myrmica fracticornis Forel
Myrmica glacialis Emery
Myrmica hamulata Weber
Myrmica incompleta Provancher
Myrmica lampra Francoeur
Myrmica latifrons Starcke
Myrmica lobifrons Pergande
Myrmica monticola Creighton
Myrmica nearctica Weber
Myrmica pinetorum Wheeler
Myrmica punctinops Francoeur
Myrmica punctiventris Roger
Myrmica quebecensis Francoeur
Myrmica rubra (Linnaeus)
Myrmica ruginodis Nylander
Myrmica rugiventris (M. R. Smith)
Myrmica semiparasitica Francoeur
Myrmica spatulata M. R. Smith
Myrmica striolagaster Cole
Myrmica tahoensis Weber
Myrmica wheeleri Weber

Myrmica wheelerorum Francoeur
Myrmica whymperi Forel
Neivamyrmex agilis Borgmeier
Neivamyrmex andrei (Emery)
Neivamyrmex baylori Watkins
Neivamyrmex californicus (Mayr)
Neivamyrmex carolinensis (Emery)
Neivamyrmex fuscipennis
   (M. R. Smith)
Neivamyrmex goyahkla Snelling &
   Snelling
Neivamyrmex graciellae (Mann)
Neivamyrmex harrisii (Haldeman)
Neivamyrmex kiowapache Snelling &
   Snelling
Neivamyrmex leonardi (Wheeler)
Neivamyrmex macropterus Borgmeier
Neivamyrmex mandibularis
   (M. R. Smith)
Neivamyrmex melanocephalus (Emery)
Neivamyrmex melshaemeri (Haldeman)
Neivamyrmex microps Borgmeier
Neivamyrmex minor (Cresson)
Neivamyrmex mojave (M. R. Smith)
Neivamyrmex moseri Watkins
Neivamyrmex ndeh Snelling &
   Snelling
Neivamyrmex nigrescens (Cresson)
Neivamyrmex nyensis Watkins
Neivamyrmex opacithorax (Emery)
Neivamyrmex pauxillus (Wheeler)
Neivamyrmex pilosus (F. Smith)
Neivamyrmex rugulosus Borgmeier
Neivamyrmex swainsonii (Shuckard)
Neivamyrmex texanus Watkins
Neivamyrmex wilsoni Snelling &
   Snelling
Nesomyrmex wilda (M. R. Smith)
Nomamyrmex esenbeckii (Westwood)
Ochetellus glaber (Mayr)
Odontomachus brunneus (Patton)
Odontomachus clarus Roger
Odontomachus haematodus (Linnaeus)
Odontomachus relictus Deyrup
   & Cover
Odontomachus ruginodis M. R. Smith
Pachycondyla chinensis (Emery)
Pachycondyla harpax (Fabricius)

Pachycondyla stigma (Fabricius)
Pachycondyla villosa (Fabricius)
Paratrechina arenivaga (Wheeler)
Paratrechina austroccidua Trager
Paratrechina bourbonica (Forel)
Paratrechina bruesii (Wheeler)
Paratrechina concinna Trager
Paratrechina faisonensis (Forel)
Paratrechina flavipes (F. Smith)
Paratrechina guatemalensis (Forel)
Paratrechina hystrix Trager
Paratrechina longicornis (Latreille)
Paratrechina parvula (Mayr)
Paratrechina phantasma Trager
Paratrechina pubens (Forel)
Paratrechina terricola (Buckley)
Paratrechina vividula (Nylander)
Paratrechina wojciki Trager
Pheidole absurda Forel
Pheidole adrianoi Naves
Pheidole artemisia Cole
Pheidole aurea Wilson
Pheidole barbata Wheeler
Pheidole bicarinata Mayr
Pheidole bureni Wilson
Pheidole californica Mayr
Pheidole carrolli Naves
Pheidole casta Wheeler
Pheidole cavigenis Wheeler
Pheidole cerebrosior Wheeler
Pheidole ceres Wheeler
Pheidole clementensis Gregg
Pheidole clydei Gregg
Pheidole cockerelli Wheeler
Pheidole coloradensis Emery
Pheidole constipata Wheeler
Pheidole crassicornis Emery
Pheidole creightoni Gregg
Pheidole davisi Wheeler
Pheidole dentata Mayr
Pheidole dentigula M. R. Smith
Pheidole desertorum Wheeler
Pheidole diversipilosa Wheeler
Pheidole elecebra (Wheeler)
Pheidole fervens (F. Smith)
Pheidole flavens Roger
Pheidole floridana Emery
Pheidole furtiva Wilson

Pheidole gilvescens Creighton
& Gregg
Pheidole grundmanni M. R. Smith
Pheidole hoplitica Wilson
Pheidole humeralis Wheeler
Pheidole hyatti Emery
Pheidole inquilina (Wheeler)
Pheidole juniperae Wilson
Pheidole lamia Wheeler
Pheidole littoralis Cole
Pheidole macclendoni Wheeler
Pheidole marcidula Wheeler
Pheidole megacephala (Fabricius)
Pheidole mera Wilson
Pheidole metallescens Emery
Pheidole micula Wheeler
Pheidole militicida Wheeler
Pheidole moerens Wheeler
Pheidole morrisii Forel
Pheidole nuculiceps Wheeler
Pheidole obscurithorax Naves
Pheidole obtusospinosa Pergande
Pheidole pacifica Wheeler
Pheidole paiute Gregg
Pheidole pelor Wilson
Pheidole perpilosa Wilson
Pheidole pilifera (Roger)
Pheidole pilosior Wilson
Pheidole pinealis Wheeler
Pheidole porcula Wheeler
Pheidole portalensis Wilson
Pheidole psammophila Creighton
& Gregg
Pheidole rhea Wheeler
Pheidole rufescens Wheeler
Pheidole rugulosa Gregg
Pheidole sciara Cole
Pheidole sciophila Wheeler
Pheidole senex Gregg
Pheidole sitiens Wilson
Pheidole soritis Wheeler
Pheidole spadonia Wheeler
Pheidole teneriffana Forel
Pheidole tepicana Pergande
Pheidole tetra Creighton
Pheidole texana Wheeler
Pheidole titanis Wheeler
Pheidole tysoni Forel

Pheidole vallicola Wheeler
Pheidole virago Wheeler
Pheidole vistana Forel
Pheidole xerophila Wheeler
Pheidole yaqui Creighton & Gregg
Plagiolepis alluaudi Emery
Platythyrea punctata (F. Smith)
Pogonomyrmex anergismus Cole
Pogonomyrmex anzensis Cole
Pogonomyrmex apache Wheeler
Pogonomyrmex badius (Latreille)
Pogonomyrmex barbatus (F. Smith)
Pogonomyrmex bicolor Cole
Pogonomyrmex bigbendensis Francke
& Merickel
Pogonomyrmex brevispinosus Cole
Pogonomyrmex californicus (Buckley)
Pogonomyrmex colei Snelling
Pogonomyrmex comanche Wheeler
Pogonomyrmex desertorum Wheeler
Pogonomyrmex huachucanus Wheeler
Pogonomyrmex imberbiculus Wheeler
Pogonomyrmex magnacanthus Cole
Pogonomyrmex maricopa Wheeler
Pogonomyrmex montanus MacKay
Pogonomyrmex occidentalis (Cresson)
Pogonomyrmex pima Wheeler
Pogonomyrmex rugosus Emery
Pogonomyrmex salinus Olsen
Pogonomyrmex subdentatus Mayr
Pogonomyrmex subnitidus Emery
Pogonomyrmex tenuisipinus Forel
Pogonomyrmex texanus Francke
& Merickel
Polyergus breviceps Emery
Polyergus lucidus Mayr
Ponera exotica M. R. Smith
Ponera pennsylvanica Buckley
Prenolepis imparis (Say)
Prionopelta antillana Forel
Proceratium californicum Cook
Proceratium chickasaw de Andrade
Proceratium compitale Ward
Proceratium crassicorne Emery
Proceratium creek de Andrade
Proceratium croceum (Roger)
Proceratium pergandei (Emery)
Proceratium silaceum Roger

Protomognathus americanus (Emery)
Pseudomyrmex apache Creighton
Pseudomyrmex caeciliae (Forel)
Pseudomyrmex cubaensis (Forel)
Pseudomyrmex ejectus (F. Smith)
Pseudomyrmex elongatus (Mayr)
Pseudomyrmex gracilis (Fabricius)
Pseudomyrmex leptosus Ward
Pseudomyrmex pallidus (F. Smith)
Pseudomyrmex seminole Ward
Pseudomyrmex simplex (F. Smith)
Pyramica abdita (Wesson & Wesson)
Pyramica angulata (M. R. Smith)
Pyramica apalachicolensis (Deyrup
   & Lubertazzi)
Pyramica archboldi (Deyrup & Cover)
Pyramica arizonica (Ward)
Pyramica bimarginata (Wesson
   & Wesson)
Pyramica boltoni Deyrup
Pyramica bunki (Brown)
Pyramica californica (Brown)
Pyramica carolinensis (Brown)
Pyramica chiricahua (Ward)
Pyramica cloydi (Pfitzer)
Pyramica clypeata (Roger)
Pyramica creightoni (M. R. Smith)
Pyramica deyrupi Bolton
Pyramica dietrichi (M. R. Smith)
Pyramica eggersi (Emery)
Pyramica filirrhina (Brown)
Pyramica filitalpa (Brown)
Pyramica gundlachi Roger
Pyramica hexamera (Brown)
Pyramica hyalina Bolton
Pyramica inopina (Deyrup & Cover)
Pyramica laevinasis (M. R. Smith)
Pyramica margaritae (Forel)
Pyramica membranifera (Emery)
Pyramica memorialis (Deyrup)
Pyramica metazytes Bolton
Pyramica missouriensis (M. R. Smith)
Pyramica ohioensis (Kennedy
   & Schramm)
Pyramica ornata (Mayr)
Pyramica pergandei (Emery)
Pyramica pilinasis (Forel)
Pyramica pulchella (Emery)

Pyramica reflexa (Wesson & Wesson)
Pyramica reliquia (Ward)
Pyramica rohweri (M. R. Smith)
Pyramica rostrata (Emery)
Pyramica talpa (Weber)
Pyramica wrayi (Brown)
Rogeria creightoni Snelling
Rogeria foreli Emery
Solenopsis abdita Thompson
Solenopsis amblychila Wheeler
Solenopsis aurea Wheeler
Solenopsis carolinensis Forel
Solenopsis catalinae Wheeler
Solenopsis corticalis Forel
Solenopsis geminata (Fabricius)
Solenopsis globularia (F. Smith)
Solenopsis invicta Buren
Solenopsis krockowi Wheeler
Solenopsis molesta (Say)
Solenopsis nickersoni Thompson
Solenopsis pergandei Forel
Solenopsis phoretica Davis & Deyrup
Solenopsis picta Emery
Solenopsis pilosula Wheeler
Solenopsis puncticeps MacKay
   & Vinson
Solenopsis richteri Forel
Solenopsis salina Wheeler
Solenopsis subterranea MacKay
   & Vinson
Solenopsis tennesseensis M. R. Smith
Solenopsis texana Emery
Solenopsis tonsa Thompson
Solenopsis truncorum Forel
Solenopsis validiuscula Emery
Solenopsis xyloni McCook
Stenamma brevicorne (Mayr)
Stenamma californicum Snelling
Stenamma chiricahua Snelling
Stenamma diecki Emery
Stenamma dyscheres Snelling
Stenamma exasperatum Snelling
Stenamma fovolocephalum
   M. R. Smith
Stenamma heathi Wheeler
Stenamma huachucanum M. R. Smith
Stenamma impar Forel
Stenamma meridionale M. R. Smith

Stenamma punctatoventre Snelling
Stenamma schmittii Wheeler
Stenamma sequoiarum Wheeler
Stenamma smithi Cole
Stenamma snellingi Bolton
Stenamma wheelerorum Snelling
Strumigenys boneti Brown
Strumigenys emmae (Emery)
Strumigenys lanuginosa Wheeler
Strumigenys louisianae Roger
Strumigenys rogeri Emery
Strumigenys silvestrii Emery
Tapinoma dimmocki (Wheeler)
Tapinoma litorale Wheeler
Tapinoma melanocephalum (Fabricius)
Tapinoma sessile (Say)
Technomyrmex difficilis Forel
Temnothorax adustus (MacKay)
Temnothorax allardycei (Mann)
Temnothorax ambiguus (Emery)
Temnothorax andersoni (MacKay)
Temnothorax andrei (Emery)
Temnothorax bestelmeyeri (MacKay)
Temnothorax bradleyi (Wheeler)
Temnothorax bristoli (MacKay)
Temnothorax carinatus (Cole)
Temnothorax chandleri (MacKay)
Temnothorax cokendolpheri (MacKay)
Temnothorax coleenae (MacKay)
Temnothorax curvispinosus (Mayr)
Temnothorax duloticus (Wesson)
Temnothorax emmae (MacKay)
Temnothorax furunculus (Wheeler)
Temnothorax gallae (M. R. Smith)
Temnothorax hispidus (Cole)
Temnothorax josephi MacKay
Temnothorax liebi (MacKay)
Temnothorax longispinosus (Roger)
Temnothorax minutissimus (M. R. Smith)
Temnothorax neomexicanus (Wheeler)
Temnothorax nevadensis (Wheeler)
Temnothorax nitens (Emery)

Temnothorax obliquicanthus (Cole)
Temnothorax obturator (Wheeler)
Temnothorax oxynodis (MacKay)
Temnothorax palustris (Deyrup & Cover)
Temnothorax pergandei (Emery)
Temnothorax politus (M. R. Smith)
Temnothorax rudis (Wheeler)
Temnothorax rugatulus (Emery)
Temnothorax schaumii (Roger)
Temnothorax schmittii (Wheeler)
Temnothorax silvestrii (Santschi)
Temnothorax smithi (Baroni Urbani)
Temnothorax stenotyle (Cole)
Temnothorax subditivus (Wheeler)
Temnothorax terrigena (Wheeler)
Temnothorax texanus Wheeler
Temnothorax torrei (Aguayo)
Temnothorax tricarinatus (Emery)
Temnothorax tuscaloosae (Wilson)
Temnothorax whitfordi MacKay
Tetramorium bicarinatum (Nylander)
Tetramorium caespitum (Linnaeus)
Tetramorium caldarium (Roger)
Tetramorium hispidum (Wheeler)
Tetramorium insolens (F. Smith)
Tetramorium lanuginosum Mayr
Tetramorium pacificum Mayr
Tetramorium simillimum (F. Smith)
Tetramorium spinosum (Pergande)
Tetramorium tsushimae Emery
Trachymyrmex arizonensis (Wheeler)
Trachymyrmex carinatus MacKay
Trachymyrmex desertorum (Wheeler)
Trachymyrmex jamaicensis (Andre)
Trachymyrmex neomexicanus Cole
Trachymyrmex nogalensis Byars
Trachymyrmex septentrionalis (McCook)
Trachymyrmex turrifex (Wheeler)
Vollenhovia emeryi Wheeler
Wasmannia auropunctata (Roger)
Xenomyrmex floridanus Emery

TERMINOLOGY

Appended below is a glossary of terms used in the keys. The vocabulary associated with ant taxonomy has been an enormous nuisance for many years, primarily for two reasons. First, a number of morphological and descriptive terms used in the literature are either incorrect or inconsistent with the practice followed in other hymenopteran groups. This has spawned controversy, confusion, and the development of rival terminologies. More importantly, taxonomic keys and species descriptions have become couched in arcane descriptive language that requires considerable effort to master and fosters the notion that ant identification is a profound, difficult, and mysterious art. No solution to this problem can be implemented here, but an effort has been made to adopt the simplest possible terminology in the keys.

Abdomen In the aculeate Hymenoptera (the female members of which bear a sting, or aculeus), the visible abdomen is not the whole story. The true first abdominal segment is permanently fused to the thorax, where it is termed the propodeum. Since ants are petiolate, the petiole is actually the true second abdominal segment; if the petiole is 2-segmented, the next segment, or postpetiole, is the true third abdominal segment. Everything beyond the petiole (or petiole plus postpetiole) comprises the gaster.

Acidopore In ants of the subfamily Formicinae, the circular, nozzle-like exit of the poison gland at the apex of the gaster; it is usually surrounded by a distinctive fringe of hairs.

Alate Winged.

Alitrunk In the aculeate Hymenoptera, it is incorrect to call the middle portion of the body the thorax. This is because, in aculeates, the middle section of the body consists of the true thorax fused with the true first abdominal segment (i.e., the propodeum). Alitrunk is a synonym of the preferred term, mesosoma.

Antennal club Refers to the last segments of the antennae (one, two, three, or four), which are conspicuously enlarged relative to the more basal segments and form a club-like apex.

Antennal insertion The condyles of the antennal scape are articulated within the two antennal insertions.

Antennal socket The cavity or depression surrounding the socket into which the antennal scape is articulated on the front of the head.

Apical At the tip or apex of a structure.

Appressed Refers to hairs that lie on the body surface and are thus parallel, or nearly so, to that surface.

Apterous Wingless.

Basal Situated at or toward the base.

Carina An elevated ridge.

Carinate Possessing one or more carinae.

Caste Members of an ant colony that are both morphologically and functionally defined (i.e., workers, female reproductives or queens, and males). There may also be subcastes, such as major and minor workers, or soldiers, etc.

Clypeus The foremost section of the head, just behind the mandibles, demarcated posteriorly by the posterior clypeal margin and anteriorly by the anterior margin of the head.

Condyle The often ball-like structure that articulates an appendage to the surface of the body, such as the basal condyle of the antennal scape.

Costate Covered with a series of close-set ridges with rounded summits.

Dealate Having formerly possessed wings, now shed; also an individual that formerly had wings.

Declivity A downward-sloping surface, such as the posterior slope of the propodeum.

Decumbent Refers to a hair or seta inclined 10 to 40 degrees from a surface.

Dentate Possessing teeth, such as the toothed margin of the mandibles.

Denticulate With many minute teeth.

Depressed Pressed downward, such as in a propodeum depressed below the margin of the promesonotum.

Dimorphic Within the caste system of an ant colony, the existence of two size classes or subcastes not connected by intermediates.

Distal Farthest away from the body, or the farthest part of a given structure, such as the tip of a wing.

Dorsal Referring to the dorsum or upper surface; the opposite is ventral.

Edentate Without teeth.

Epinotum See *propodeum.*

Erect Refers to a hair that stands straight up, or nearly so, from the body surface.

Facet An ommatidium, one of the units of the compound eye.

Falcate Sickle-shaped or saber-shaped.

Fovea A large, deep pit on the body surface.

Foveate A body surface covered with foveae.

Frontal carinae A pair of subparallel or posteriorly divergent carinae, or ridges, are located behind the clypeus and between the antennal sockets. Laterally, they frequently develop into lobes that may partially or entirely overlap the antennal sockets.

Frontal triangle A triangular area, demarcated by grooves, that lies immediately above the posterior margin of the clypeus and between the antennal sockets. Not apparent in some taxa.

Funiculus All of the antenna beyond the first segment, or scape. Sometimes unwisely referred to as the flagellum.

Gaster The last major body part of an ant, the abdomen, which follows the petiole (or petiole plus postpetiole).

Gena That area of the side of the head that lies between the compound eye and the margin that turns mesad (toward the midline of the body) to form the gula, or ventral surface of the head. It is not, as stated incorrectly by Hölldobler and Wilson (1990), "the area be-tween one of the compound eyes and the nearest antennal insertion."

Glabrous There are two different usages for this term. In the strict sense, it refers to a body surface that lacks hairs. In the broad sense, the term commonly means smooth and shiny (e.g., Hölldobler and Wilson, 1990).

Gula The central portion of the posterior surface of the head that lies between the mouthparts and the foramen magnum. See *hypostoma*.

Head length With the head in frontal view, the maximum length between the uppermost portion of the vertex and the lowermost portion of the clypeus.

Head width With the head in frontal view, the maximum visible width across the head, exclusive of the compound eyes.

Humerus With the mesosoma in dorsal view, the anterolateral corner or angle of the pronotum. May also be referred to as humeral angles. The plural is *humeri*.

Hypostoma The anteroventral region of the head forming the posterior border of the oral cavity and extending laterally to the base of the mandibles. The border between the oral cavity and the underside of the head capsule is marked by the hypostomal carina. In older literature, this region is sometimes incorrectly referred to as the gula.

Hypostomal teeth The pair or two pairs of teeth or lobes present symmetrically on either side of the midpoint of the hypostomal carina. Present in *Dolichoderus* and *Pheidole* soldiers.

Impressed Indented or pressed in, such as in an impressed suture.

Labial palps or palpi The segmented appendages of the mouthparts that arise from the labium, or lower lip; they consist of up to four segments. See *palp count*.

Labral lobes Sometimes the lower margin of the labrum (see below) is very deeply divided to form a pair of slender lobes, especially in the tribe Dacetini (Myrmicinae).

Labrum The upper lip of insects, a movable flap attached to the lower margin of the clypeus and often folded back to cover the palpi and tongue.

Lamella A thin, plate-like process or ridge, often more or less translucent.

Major worker A member of the largest subcaste of worker ants; sometimes referred to as a soldier because a major worker is often specialized for defense, but may also be specialized to crush seeds, as in species of *Pheidole*.

Malar area The portion of the side of the face that lies between the lower margin of the compound eye and the base of the mandible. In older literature, sometimes referred to as the cheeks.

Marginate Used to describe the condition in which the edge of an area, such as the dorsum of the pronotum, is marked by a sharp angle, ridge, or flange.

Maxillary palpi Jointed appendages originating from the maxilla. The maximum number of segments is six, but the number of segments will be fewer or none in different taxa. See *palp count.*

Mesonotum The middle sclerite (plate) on the mesosomal dorsum, between the pronotum and the propodeum.

Mesosoma The middle of the three parts of the insect body; also variously known as alitrunk, trunk, or truncus. As the hymenopteran mesosoma consists of the true thorax plus the first true abdominal segment, the use of the term thorax in the higher Hymenoptera is incorrect and should be avoided.

Metanotal groove or impression A transverse impression or groove separating the mesonotum from the propodeum on the mesosomal dorsum.

Metapleural gland A gland peculiar to the ants, located at the posteroventral angle of the metapleuron (the lateral area of the mesosoma above the hind coax), which produces antibiotic substances.

Microreticulate Fine sculpturing that forms a net-like pattern.

Minor worker A member of the smallest of the worker subcastes.

Node As used in ant taxonomy, a vertically projecting extension of the petiole or postpetiole that may be rounded, angulate, toothed, or thin and scale-like. See *petiolar scale.*

Occipital lobes An obsolete and incorrect term; see *vertex.*

Occiput A frequently used term referring to the rear of the head of the ant. The use of this term, however, is incorrect; the appropriate term is vertex.

Ocellus Any one of the three simple eyes of the adult head, centrally located on or near the vertex and arranged in a triangle.

Palp count [#,#] indicates the number of segments in the maxillary and labial palps, respectively. For example, a palp count of 6,4 indicates six maxillary segments and four labial.

Pectinate Comb-like or bearing a comb.

Peduncle The stalk anterior to the node of the petiole.

Petiolar scale In dolichoderine and formicine ants, refers to the dorsal node of the petiole; the term scale may be used when the node is compressed from front to back and is thus rather thin and scale-like when viewed in profile.

Petiole The waist between the mesosoma and the metasoma or gaster; if the waist is 2-segmented, then the first of these is the petiole (pt) and the second is the postpetiole (ppt).

Pilosity The longer, stouter hairs or setae, which stand out above the shorter and finer hairs constituting the pubescence.

Polymorphic Refers to the coexistence of two or more subcastes within the same caste, often serving different functions and connected through a gradual series of intermediates in a nonisometric growth curve; individuals of distinctly different proportions occur at the extreme ends of the variation range. Also *polymorphism*.

Postpetiole The second of two segments forming the waist in certain ant groups. See *petiole*.

Promesonotum The pronotum and mesonotum are sometimes fused to form one structure called the promesonotum.

Pronotum The first (anterior) tergite on the mesosomal dorsum.

Propodeum In the higher Hymenoptera, the true first abdominal segment, fused to the true thorax to form the mesosoma. In ant taxonomy, it has been called the epinotum, a redundant and obsolete term.

Psammophore The fringe of long hairs on the posterior surface of the head (found in a few genera of desert ants).

Pubescence The fine, short hairs that usually form a second layer beneath the longer, coarser pilosity. The pubescence is commonly appressed, and less commonly is suberect.

Pygidium Correctly, the last visible tergite of the gaster.

Replete A worker ant whose crop is greatly distended with liquid food and functions as a living reservoir; this food is made available to other colony members by regurgitation.

Ruga A wrinkle on the body surface.

Rugoreticulate Refers to a surface with irregular, coarse rugae that form a network. Differs from reticulate by being coarser and more irregular.

Rugose Refers to a surface with multiple rugae that are approximately parallel.

Scape The first antennal segment, articulated to the head via the antennal socket. In female ants, this segment is enormously elongated compared to the succeeding segments of the antenna.

Scrobe A groove, often marginate, into which an appendage may be folded.

Serrate Bearing fine teeth along the edge, in a manner similar to a saw blade.

Shagreened Refers to a surface covered with fine, close-set roughness, in a manner similar to a sharkskin.

Soldier A worker subcaste specialized for colony defense, often with an enlarged head and/or mandibles.

Spur A spine-like appendage at the apex of the tibia; often paired.

Squamate Refers to a broad, flattened, scale-like hair.

Sternite In ant taxonomy, most commonly used to refer to the ventral sclerite (plate) of an abdominal segment.

Stria A fine, impressed line on the body surface; usually many striae occur together, resulting in a striate surface.

Suberect Refers to a hair that stands at an angle of about 45 degrees from the body surface.

Subgenus A distinctive, presumably monophyletic, group of species within a genus. Rarely, a single species may be so distinctive that it is considered a separate subgenus. Subgenera are seldom used now; groups of closely related species are referred to as a species group. Yet subgeneric names are still used to refer to groups long known in the literature by those names.

Sulcus A deep furrow or groove.

Taxon A taxonomic entity, such as a species or genus. The plural is *taxa*.

Tergite In ant taxonomy, most commonly used to refer to the dorsal sclerite (plate) of an abdominal segment.

Tubercles Short, thick, usually blunt spines or pimple-like structures.

Tuberculate Refers to a surface that bears a number of tubercles.

Ventral Refers to the lower surface of a body part.

Vertex The top of the head, often incorrectly called the occiput.

Waist A collective term for the one or two segments separating the propodeum from the gaster.

Worker The laboring caste of ant colonies that performs all routine nest activities except reproduction.

IDENTIFICATION
REFERENCES

Acanthostichus

MacKay, W. P. (1996) A revision of the ant genus *Acanthostichus* (Hymenoptera: Formicidae). *Sociobiology,* 27(2), 129–179.

Acropyga

Lapolla, J. S. (2004) Acropyga *(Hymenoptera: Formicidae) of the world.* Contributions of the American Entomological Institute, 33(3).

Amblyopone

Ward, P. S. (1988) Mesic elements in the western Nearctic ant fauna: taxonomic and biological notes on *Amblyopone, Proceratium,* and *Smithistruma* (Hymenoptera: Formicidae). *Journal of the Kansas Entomological Society,* 61, 102–124.

Anochetus

Brown, W. L., Jr. (1978) Contributions toward a reclassification of the Formicidae. Part VI. Ponerinae, tribe Ponerini, subtribe Odontomachiti. Section B. Genus *Anochetus* and bibliography. *Studia Entomologica,* 20, 549–638.

Aphaenogaster

Umphrey, G. J. (1996) Morphometric discrimination among sibling species in the *fulva-rudis-texana* complex of the ant genus *Aphaenogaster* (Hymenoptera: Formicidae). *Canadian Journal of Zoology,* 74, 528–559.

Camponotus

Creighton, W. S. (1950) The ants of North America. *Bulletin of the Museum of Comparative Zoology*, 104, 1–585.

Snelling, R. R. (1988) Taxonomic notes on Nearctic species of *Camponotus*, subgenus *Myrmentoma* (Hymenoptera: Formicidae). In *Advances in Myrmecology*, edited by J. C. Trager. Leiden, The Netherlands: E. J. Brill. 55–78.

Cardiocondyla

Seifert, B. (2003) The ant genus *Cardiocondyla* (Insecta: Hymenoptera: Formicidae)—a taxonomic revision of the *C. elegans, C. bulgarica, C. batesii, C. nuda, C. shuckardi, C. stambuloffii, C. wroughtonii, C. emeryi,* and *C. minutior* species groups. *Annalen des Naturhistorischen Museums in Wien. B, Botanik und Zoologie*, 104, 203–338.

Carebara

Fernandez, F. (2004) The American species of the myrmecine ant genus *Carebara* Westwood (Hymenoptera: Formicidae). *Caldasia*, 26(1), 191–238.

Cephalotes

De Andrade, M. L., and C. Baroni Urbani. (1999) Diversity and adaptation in the ant genus *Cephalotes*, past and present. *Stuttgarter Beiträge zur Naturkunde Serie B (Geologie und Paläontologie)*, 271, 1–889.

Crematogaster

Buren, W. F. (1968) A review of the species of *Crematogaster*, sensu stricto, in North America (Hymenoptera, Formicidae). Part II. Descriptions of new species. *Journal of the Georgia Entomological Society*, 3, 91–121.

Longino, J. T. (2003) The *Crematogaster* (Hymenoptera, Formicidae, Myrmicinae) of Costa Rica. *Zootaxa*, 151, 1–150.

Cryptopone

Brown, W. L., Jr. (1963) Characters and synonymies among the genera of ants. Part III. Some members of the tribe Ponerini (Ponerinae, Formicidae). *Breviora*, 190, 1–10.

Cyphomyrmex

Kempf, W. W. (1964) A revision of the Neotropical fungus-growing ants of the genus *Cyphomyrmex* Mayr. Part I: Group of *strigatus* Mayr (Hymenoptera, Formicidae). *Studia Entomologica,* 7, 1–44.

Snelling, R. R., and J. T. Longino. (1992) Revisionary notes on the fungus-growing ants of the genus *Cyphomyrmex, rimosus* group (Hymenoptera: Formicidae: Attini). In *Insects of Panama and Mesoamerica: selected studies,* edited by D. Quintero and A. Aiello. Oxford: Oxford University Press. 479–494.

Dolichoderus

Creighton, W. S. (1950) The ants of North America. *Bulletin of the Museum of Comparative Zoology,* 104, 1–585.

MacKay, W. P. (1993) A review of the New World ants of the genus *Dolichoderus* (Hymenoptera: Formicidae). *Sociobiology,* 22, 1–148.

Dolopomyrmex

Cover, S. P., and M. Deyrup. (2007) A new ant genus from the southwestern United States. In Snelling, R. R., B. L. Fisher, and P. S. Ward, eds. *Advances in ant systematics (Hymenoptera: Formicidae): Homage to E. O. Wilson—50 years of contributions.* Memoirs of the American Entomological Institute, 80, 100–112.

Dorymyrmex

Snelling, R. R. (1995) Systematics of Nearctic ants of the genus *Dorymyrmex* (Hymenoptera: Formicidae). *Contributions in Science* (Natural History Museum of Los Angeles County), 454, 1–14.

Trager, J. C. (1988) A revision of *Conomyrma* (Hymenoptera: Formicidae) from the southeastern United States, especially Florida, with keys to the species. *Florida Entomologist,* 71, 11–29.

Forelius

Cuezzo, F. (2000) Revisión del género *Forelius* (Hymenoptera: Formicidae: Dolichoderinae). *Sociobiology,* 35, 197–275.

Formica

Agosti, D., and B. Bolton. (1990) New characters to differentiate the ant genera *Lasius* F. and *Formica* L. (Hymenoptera: Formicidae). *Entomologist's Gazette,* 41, 149–156.

Creighton, W. S. (1950) The ants of North America. *Bulletin of the Museum of Comparative Zoology,* 104, 1–585.

Francoeur, A. (1973) Révision taxonomique des espèces néarctiques du groupe *fusca,* genre *Formica* (Formicidae, Hymenoptera). *Mémoires de la Société Entomologique du Québec,* 3, 1–316.

Snelling, R. R., and W. F. Buren. (1985) Description of a new species of slave-making ant in the *Formica sanguinea* group (Hymenoptera: Formicidae). *Great Lakes Entomologist,* 18, 69–78.

Trager, J. C., J. A. MacGown, and M. D. Trager. (2007) Revision of the Nearctic endemic *Formica pallidefulva* group. In Snelling, R. R., B. L. Fisher, and P. S. Ward, eds. *Advances in ant systematics (Hymenoptera: Formicidae): Homage to E. O. Wilson—50 years of contributions.* Memoirs of the American Entomological Institute, 80, 610–636.

Formicoxenus

Francoeur, A., R. Loiselle, and A. Buschinger. (1985) Biosystématique de la tribu Leptothoracini (Formicidae, Hymenoptera). 1. Le genre *Formicoxenus* dans la région holarctique. *Naturaliste Canadien* (Québec), 112, 343–403.

Gnamptogenys

Lattke, J. E. (1995) Revision of the ant genus *Gnamptogenys* in the New World (Hymenoptera: Formicidae). *Journal of Hymenoptera Research,* 4, 137–193.

Labidus

Watkins, J. F., II. (1985) The identification and distribution of the army ants of the United States of America (Hymenoptera, Formicidae, Ecitoninae). *Journal of the Kansas Entomological Society,* 58, 479–502.

Lasius

Wilson, E. O. (1955) A monographic revision of the ant genus *Lasius. Bulletin of the Museum of Comparative Zoology,* 113, 1–201.

Wing, M. W. (1968) Taxonomic revision of the Nearctic genus *Acanthomyops* (Hymenoptera: Formicidae). *Memoirs of the Cornell University Agricultural Experiment Station,* 405, 1–173.

Leptogenys

Trager, J. C., and C. Johnson. (1988) The ant genus *Leptogenys* (Hymenoptera: Formicidae, Ponerinae) in the United States. In *Advances in Myrmecology,* edited by J. C. Trager. Leiden, The Netherlands: E. J. Brill. 29–34.

Linepithema

Wild, A. L. (2004) Taxonomy and distribution of the Argentine Ant *Linepithema humile* (Hymenoptera: Formicidae). *Annals of the Entomological Society of America,* 97, 1204–1215.

Liometopum

Creighton, W. S. (1950). The ants of North America. *Bulletin of the Museum of Comparative Zoology,* 104, 1–585.

Manica

Wheeler, G. C., and J. Wheeler. (1970) The natural history of *Manica* (Hymenoptera: Formicidae). *Journal of the Kansas Entomological Society,* 43, 129–162.

Messor

Johnson, R. A. (2000) Seed-harvester ants (Hymenoptera: Formicidae) of North America: an overview of ecology and biogeography. *Sociobiology,* 36, 89–122 and 83–88.

Johnson, R. A. (2001) Biogeography and community structure of North American seed-harvesting ants. *Annual Review of Entomology,* 46, 1–29.

Smith, M. R. (1956) A key to the workers of *Veromessor* Forel of the United States and the description of a new subspecies (Hymenoptera, Formicidae). *Pan-Pacific Entomologist,* 32, 36–38.

Monomorium

DuBois, M. B. (1986) A revision of the native New World species of the ant genus *Monomorium* (*minimum* group) (Hymenoptera: Formicidae). *University of Kansas Science Bulletin,* 53, 65–119.

Mycetosoritis

Wheeler, W. M. (1907) The fungus-growing ants of North America. *Bulletin of the American Museum of Natural History,* 23, 669–807.

Myrmecocystus

Snelling, R. R. (1976) A revision of the honey ants, genus *Myrmecocystus* (Hymenoptera: Formicidae). *Natural History Museum of Los Angeles County Science Bulletin,* 24, 1–163.

Snelling, R. R. (1982) A revision of the honey ants, genus *Myrmeco-cystus*, first supplement (Hymenoptera: Formicidae). *Bulletin of the Southern California Academy of Sciences,* 81, 69–86.

Francoeur, A. (2007). The *Myrmica punctiventris* and *M. crassirugis* species groups in the Nearctic region. In Snelling, R. R., B. L. Fisher, and P. S. Ward, eds. *Advances in ant systematics (Hymenoptera: Formicidae): Homage to E. O. Wilson—50 years of contributions.* Memoirs of the American Entomological Institute, 80, 153–185.

Neivamyrmex

Snelling, G. C., and R. R. Snelling. (2007) New synonymy, new species, new keys to *Neivamyrmex* army ants of the United States. In Snelling, R. R., B. L. Fisher, and P. S. Ward, eds. *Advances in ant systematics (Hymenoptera: Formicidae): Homage to E. O. Wilson—50 years of contributions.* Memoirs of the American Entomological Institute, 80, 459–550.

Nexomyrmex

Kempf, W. W. (1959) A synopsis of the New World species belonging to the *Nesomyrmex*-group of the ant genus *Leptothorax* Mayr (Hymenoptera: Formicidae). *Studia Entomologica,* 2, 391–432.

Nomamyrmex

Watkins, J. F., II. (1985) The identification and distribution of the army ants of the United States of America (Hymenoptera, Formicidae, Ecitoninae). *Journal of the Kansas Entomological Society,* 58, 479–502.

Odontomachus

Deyrup, M., and S. Cover. (2004) A new species of *Odontomachus* ant (Hymenoptera: Formicidae) from inland ridges of Florida, with a key to *Odontomachus* of the United States. *Florida Entomologist,* 87(2), 136–144.

Paratrechina

Trager, J. C. (1984) A revision of the genus *Paratrechina* (Hymenoptera: Formicidae) of the continental United States. *Sociobiology,* 9, 49–162.

Pheidole

Gregg, R. E. (1959 ["1958"]) Key to the species of *Pheidole* (Hymenoptera: Formicidae) in the United States. *Journal of the New York Entomological Society,* 66, 7–48.

Wilson, E. O. (2003) Pheidole *in the New World. A dominant, hyperdiverse ant genus.* Cambridge: Harvard University Press.

Platythyrea

Brown, W. L., Jr. (1975) Contributions toward a reclassification of the Formicidae. V. Ponerinae, tribes Platythyreini, Cerapachyini, Cylindromyrmecini, Acanthostichini, and Aenictogitini. Search, Agriculture (Ithaca, N.Y.), 5, 1–116.

Pogonomyrmex

Cole, A. C., Jr. (1968) Pogonomyrmex *harvester ants. A study of the genus in North America.* Knoxville: University of Tennessee Press.

Taber, S. W. (1998) *The world of harvester ants.* College Station, Tex.: Texas A&M University Press.

Ponera

Taylor, R. W. (1967) A monographic revision of the ant genus *Ponera* Latreille (Hymenoptera: Formicidae). *Pacific Insects Monograph,* 13, 1–112.

Prionopelta

Brown, W. L., Jr. (1960) Contributions toward a reclassification of the Formicidae. III. Tribe Amblyoponini (Hymenoptera). *Bulletin of the Museum of Comparative Zoology,* 122, 143–230.

Proceratium

Baroni Urbani, C., and M. L. de Andrade. (2003) The ant genus *Proceratium* in the extant and fossil record (Hymenoptera: Formicidae). *Museo Regionale di Scienze Naturali Monografie* (Turin), 36, 1–492.

Pseudomyrmex

Ward, P. S. (1985) The Nearctic species of the genus *Pseudomyrmex* (Hymenoptera: Formicidae). *Quaestiones Entomologicae,* 21, 209–246.

Pyramica

Bolton, B. (1999) Ant genera of the tribe Dacetonini (Hymenoptera: Formicidae). *Journal of Natural History,* 33, 1639–1689.

Bolton, B. (2000) *The ant tribe* Dacetini. Memoirs of the American Entomological Institute, 65.

Rogeria

Kugler, C. (1994) A revision of the ant genus *Rogeria* with description of the sting apparatus (Hymenoptera: Formicidae). *Journal of Hymenoptera Research,* 3, 17–89.

Solenopsis

Creighton, W. S. (1950) The ants of North America. *Bulletin of the Museum of Comparative Zoology,* 104, 1–585.

Thompson, C. R. (1989a) Scientific notes: rediscovered species and revised key to the Florida thief ants (Hymenoptera: Formicidae). *Florida Entomologist,* 72, 697–698.

Thompson, C. R. (1989b) The thief ants, *Solenopsis molesta* group, of Florida (Hymenoptera: Formicidae). *Florida Entomologist,* 72, 268–283.

Trager, J. C. (1991) A revision of the fire ants, *Solenopsis geminata* group (Hymenoptera: Formicidae: Myrmicinae). *Journal of the New York Entomological Society,* 99, 141–198.

Stenamma

Smith, M. R. (1957) Revision of the genus *Stenamma* Westwood in America north of Mexico (Hymenoptera, Formicidae). *American Midland Naturalist,* 57, 133–174.

Snelling, R. R. (1973) Studies on California ants. 7. The genus *Stenamma* (Hymenoptera: Formicidae). *Contributions in Science* (Natural History Museum of Los Angeles County), 245, 1–38.

Strumigenys

Bolton, B. (2000) *The ant tribe* Dacetini. Memoirs of the American Entomological Institute, 65.

Temnothorax

MacKay, W. P. (2000) A review of the New World ants of the subgenus *Myrafant* (genus *Leptothorax*) (Hymenoptera: Formicidae). *Sociobiology,* 36, 265–444.

Tetramorium

Bolton, B. (1979) The ant tribe Tetramoriini (Hymenoptera: Formicidae). The genus *Tetramorium* Mayr in the Malagasy region and in the New World. *Bulletin of the British Museum (Natural History), Entomology,* 38, 129–181.

GENERAL REFERENCES

Agosti, D., and B. Bolton. (1990) New characters to differentiate the ant genera *Lasius* F. and *Formica* L. (Hymenoptera: Formicidae). *Entomologist's Gazette,* 41, 149–156.

Baroni Urbani, C., and M. L. de Andrade. (2003) The ant genus *Proceratium* in the extant and fossil record (Hymenoptera: Formicidae). *Museo Regionale di Scienze Naturali Monografie* (Turin), 36, 1–492.

Bolton, B. (1979) The ant tribe Tetramoriini (Hymenoptera: Formicidae). The genus *Tetramorium* Mayr in the Malagasy region and in the New World. *Bulletin of the British Museum (Natural History), Entomology,* 38, 129–181.

Bolton, B. (1994) *Identification guide to the ant genera of the world.* Cambridge: Harvard University Press.

Bolton, B. (1999) Ant genera of the tribe Dacetonini (Hymenoptera: Formicidae). *Journal of Natural History,* 33, 1639–1689.

Bolton, B. (2000) *The ant tribe* Dacetini. Memoirs of the American Entomological Institute, 65, 1–1028.

Bolton, B. (2003) *Synopsis and classification of* Formicidae. Memoirs of the American Entomological Institute, 71.

Brown, W. L., Jr. (1960) Contributions toward a reclassification of the Formicidae. III. Tribe Amblyoponini (Hymenoptera). *Bulletin of the Museum of Comparative Zoology,* 122, 143–230.

Brown, W. L., Jr. (1976) Contributions toward a reclassification of the Formicidae. Part VI. Ponerinae, tribe Ponerini, subtribe Odontomachiti. Section A. Introduction. Subtribal characters. Genus *Odontomachus. Studia Entomologica,* 19, 67–171.

Brown, W. L., Jr. (1978) Contributions toward a reclassification of the Formicidae. Part VI. Ponerinae, tribe Ponerini, subtribe Odontomachiti. Section B. Genus *Anochetus* and bibliography. *Studia Entomologica,* 20, 549–638.

Buren, W. F. (1968) A review of the species of *Crematogaster,* sensu stricto, in North America (Hymenoptera, Formicidae). Part II. Descriptions of new species. *Journal of the Georgia Entomological Society,* 3, 91–121.

Cole, A. C., Jr. (1968) Pogonomyrmex *harvester ants. A study of the genus in North America.* Knoxville: University of Tennessee Press.

Cover, S. (1990) A key to the ant genera of North America and northern and central Mexico, known as the Nearctic region. In *The Ants,* Bert Hölldobler and Edward O. Wilson. Cambridge: Harvard University Press, 63–69.

Cover, S. P., and M. Deyrup. (2007) A new ant genus from the southwestern United States. In Snelling, R. R., B. L. Fisher, and P. S. Ward, eds. *Advances in ant systematics (Hymenoptera: Formicidae): Homage to E. O. Wilson—50 years of contributions.* Memoirs of the American Entomological Institute, 80, 89–99.

Creighton, W. S. (1950) The ants of North America. *Bulletin of the Museum of Comparative Zoology,* 104, 1–585.

Cuezzo, F. (2000) Revisión del género *Forelius* (Hymenoptera: Formicidae: Dolichoderinae). *Sociobiology,* 35, 197–275.

De Andrade, M. L., and C. Baroni Urbani. (1999) Diversity and adaptation in the ant genus *Cephalotes,* past and present. *Stuttgarter Beiträge zur Naturkunde Serie B (Geologie und Paläontologie),* 271, 1–889.

Deyrup, M., and S. Cover. (2004a) A new species of *Odontomachus* ant (Hymenoptera: Formicidae) from inland ridges of Florida, with a key to *Odontomachus* of the United States. *Florida Entomologist,* 87(2), 136–144.

Deyrup, M., and S. Cover. (2004b) A new species of the ant genus *Leptothorax* from Florida, with a key to the *Leptothorax* of the southeast (Hymenoptera: Formicidae). *Florida Entomologist,* 87, 51–59.

Deyrup, M. and S. P. Cover. (2007) A new species of *Crematogaster* from the pinelands of the southeastern United States. In Snelling, R. R., B. L. Fisher, and P. S. Ward, eds. *Advances in ant systematics (Hymenoptera: Formicidae): Homage to E. O. Wilson—50 years of contributions.* Memoirs of the American Entomological Institute, 80, 100–112.

DuBois, M. B. (1986) A revision of the native New World species of the ant genus *Monomorium* (*minimum* group) (Hymenoptera: Formicidae). *University of Kansas Science Bulletin,* 53, 65–119.

Fernandez, F. (2004) The American species of the myrmicine ant genus *Carebara* Westwood (Hymenoptera: Formicidae). *Caldasia,* 26(1), 191–23.

Francoeur, A. (1973) Révision taxonomique des espèces néarctiques du groupe *fusca,* genre *Formica* (Formicidae, Hymenoptera). *Mémoires de la Société Entomologique du Québec,* 3, 1–316.

Francoeur, A. (2007). The *Myrmica punctiventris* and *M. crassirugis* species groups in the Nearctic region. In Snelling, R. R., B. L. Fisher, and P. S. Ward, eds. *Advances in ant systematics (Hymenoptera: Formicidae): Homage to E. O. Wilson — 50 years of contributions.* Memoirs of the American Entomological Institute, 80, 153–185.

Francoeur, A., R. Loiselle, and A. Buschinger. (1985) Biosystématique de la tribu Leptothoracini (Formicidae, Hymenoptera). 1. Le genre *Formicoxenus* dans la région holarctique. *Naturaliste Canadien* (Québec), 112, 343–403.

Gregg, R. E. (1959 ["1958"]) Key to the species of *Pheidole* (Hymenoptera: Formicidae) in the United States. *Journal of the New York Entomological Society,* 66, 7–48.

Hölldobler, B., and E. O. Wilson. (1986) Ecology and behavior of the primitive, cryptobiotic ant *Prionopelta amabilis* (Hymenoptera: Formicidae). *Insectes Sociaux,* 33, 45–58.

Johnson, R. A. (2000) Seed-harvester ants (Hymenoptera: Formicidae) of North America: an overview of ecology and biogeography. *Sociobiology,* 36, 83–88 and 89–122.

Johnson, R. A. (2001) Biogeography and community structure of North American seed-harvesting ants. *Annual Review of Entomology,* 46, 1–29.

Kempf, W. W. (1959) A synopsis of the New World species belonging to the *Nesomyrmex*-group of the ant genus *Leptothorax* Mayr (Hymenoptera: Formicidae). *Studia Entomologica,* 2, 391–432.

Kempf, W. W. (1964) A revision of the Neotropical fungus-growing ants of the genus *Cyphomyrmex* Mayr. Part I: Group of *strigatus* Mayr (Hymenoptera, Formicidae). *Studia Entomologica,* 7, 1–44.

Kugler, C. (1994) A revision of the ant genus *Rogeria* with description of the sting apparatus (Hymenoptera: Formicidae). *Journal of Hymenoptera Research,* 3, 17–89.

Lapolla, J. S. (2004) Acropyga *(Hymenoptera: Formicidae) of the world.* Contributions of the American Entomological Institute, 33(3).

Longino, J. T. (2003) The *Crematogaster* (Hymenoptera, Formicidae, Myrmicinae) of Costa Rica. *Zootaxa,* 151, 1–150.

MacKay, W. P. (1993) A review of the New World ants of the genus *Dolichoderus* (Hymenoptera: Formicidae). *Sociobiology,* 22, 1–148.

MacKay, W. P. (1996) A revision of the ant genus *Acanthostichus* (Hymenoptera: Formicidae). *Sociobiology,* 27(2), 129–179.

MacKay, W. P. (2000) A review of the New World ants of the subgenus *Myrafant* (genus *Leptothorax*) (Hymenoptera: Formicidae). *Sociobiology,* 36, 265–444.

Schilder, K., J. Heinze, and B. Hölldobler. (1999) Colony structure and reproduction in the thelytokous parthenogenetic ant *Platythyrea punctata* (F. Smith) (Hymenoptera: Formicidae). *Insectes Sociaux,* 46, 150–158.

Seifert, B. (2003) The ant genus *Cardiocondyla* (Insecta: Hymenoptera: Formicidae)—a taxonomic revision of the *C. elegans, C. bulgarica, C. batesii, C. nuda, C. shuckardi, C. stambuloffii, C. wroughtonii, C. emeryi,* and *C. minutior* species groups. *Annalen des Naturhistorischen Museums in Wien. B, Botanik und Zoologie,* 104, 203–338.

Smith, M. R. (1956) A key to the workers of *Veromessor* Forel of the United States and the description of a new subspecies (Hymenoptera, Formicidae). *Pan-Pacific Entomologist,* 32, 36–38.

Smith, M. R. (1957) Revision of the genus *Stenamma* Westwood in America north of Mexico (Hymenoptera, Formicidae). *American Midland Naturalist,* 57, 133–174.

Snelling, G. C., and R. R. Snelling. (2007) New synonymy, new species, new keys to *Neivamyrmex* army ants of the United States. In Snelling, R. R., B. L. Fisher, and P. S. Ward, eds. *Advances in ant systematics (Hymenoptera: Formicidae): Homage to E. O. Wilson—50 years of contributions.* Memoirs of the American Entomological Institute, 80, 459–550.

Snelling, R. R. (1973) Studies on California ants. 7. The genus *Stenamma* (Hymenoptera: Formicidae). *Contributions in Science* (Natural History Museum of Los Angeles County), 245, 1–38.

Snelling, R. R. (1976) A revision of the honey ants, genus *Myrmecocystus* (Hymenoptera: Formicidae). *Natural History Museum of Los Angeles County Science Bulletin,* 24, 1–163.

Snelling, R. R. (1982) A revision of the honey ants, genus *Myrmecocystus,* first supplement (Hymenoptera: Formicidae). *Bulletin of the Southern California Academy of Sciences,* 81, 69–86.

Snelling, R. R. (1988) Taxonomic notes on Nearctic species of *Camponotus,* subgenus *Myrmentoma* (Hymenoptera: Formicidae). In

Advances in Myrmecology, edited by J. C. Trager. Leiden, The Netherlands: E. J. Brill. 55–78.

Snelling, R. R. (1995) Systematics of Nearctic ants of the genus *Dorymyrmex* (Hymenoptera: Formicidae). *Contributions in Science* (Natural History Museum of Los Angeles County), 454, 1–14.

Snelling, R. R., and W. F. Buren. (1985) Description of a new species of slave-making ant in the *Formica sanguinea* group (Hymenoptera: Formicidae). *Great Lakes Entomologist,* 18, 69–78.

Snelling, R. R., and J. T. Longino. (1992) Revisionary notes on the fungus-growing ants of the genus *Cyphomyrmex, rimosus* group (Hymenoptera: Formicidae: Attini). In *Insects of Panama and Mesoamerica: selected studies,* edited by D. Quintero and A. Aiello. Oxford: Oxford University Press. 479–494.

Taber, S. W. (1998) *The world of harvester ants.* College Station, Tex.: Texas A&M University Press.

Taylor, R. W. (1967) A monographic revision of the ant genus *Ponera* Latreille (Hymenoptera: Formicidae). *Pacific Insects Monograph,* 13, 1–112.

Thompson, C. R. (1989a) Scientific notes: rediscovered species and revised key to the Florida thief ants (Hymenoptera: Formicidae). *Florida Entomologist,* 72, 697–698.

Thompson, C. R. (1989b) The thief ants, *Solenopsis molesta* group, of Florida (Hymenoptera: Formicidae). *Florida Entomologist,* 72, 268–283.

Trager, J. C. (1984) A revision of the genus *Paratrechina* (Hymenoptera: Formicidae) of the continental United States. *Sociobiology,* 9, 49–162.

Trager, J. C. (1988) A revision of *Conomyrma* (Hymenoptera: Formicidae) from the southeastern United States, especially Florida, with keys to the species. *Florida Entomologist,* 71, 11–29.

Trager, J. C. (1991) A revision of the fire ants, *Solenopsis geminata* group (Hymenoptera: Formicidae: Myrmicinae). *Journal of the New York Entomological Society,* 99, 141–198.

Trager, J. C., and C. Johnson. (1988) The ant genus *Leptogenys* (Hymenoptera: Formicidae, Ponerinae) in the United States. In *Advances in Myrmecology,* edited by J. C. Trager. Leiden, The Netherlands: E. J. Brill. 29–34.

Trager, J. C., J. A. MacGown, and M. D. Trager. (2007) Revision of the Nearctic endemic *Formica pallidefulva* group. In Snelling, R. R., B. L. Fisher, and P. S. Ward, eds. *Advances in ant systematics (Hy-*

menoptera: Formicidae): Homage to E. O. Wilson—50 years of contributions. Memoirs of the American Entomological Institute, 80, 610–636.

Umphrey, G. J. (1996) Morphometric discrimination among sibling species in the *fulva-rudis-texana* complex of the ant genus *Aphaenogaster* (Hymenoptera: Formicidae). *Canadian Journal of Zoology,* 74, 528–559.

Ward, P. S. (1985) The Nearctic species of the genus *Pseudomyrmex* (Hymenoptera: Formicidae). *Quaestiones Entomologicae,* 21, 209–246.

Ward, P. S. (1988) Mesic elements in the western Nearctic ant fauna: taxonomic and biological notes on *Amblyopone, Proceratium,* and *Smithistruma* (Hymenoptera: Formicidae). *Journal of the Kansas Entomological Society,* 61, 102–124.

Watkins, J. F., II. (1985) The identification and distribution of the army ants of the United States of America (Hymenoptera, Formicidae, Ecitoninae). *Journal of the Kansas Entomological Society,* 58, 479–502.

Wheeler, G. C., and J. Wheeler. (1970) The natural history of *Manica* (Hymenoptera: Formicidae). *Journal of the Kansas Entomological Society,* 43, 129–162.

Wheeler, W. M. (1907) The fungus-growing ants of North America. *Bulletin of the American Museum of Natural History,* 23, 669–807.

Wilson, E. O. (1955) A monographic revision of the ant genus *Lasius. Bulletin of the Museum of Comparative Zoology,* 113, 1–201.

Wilson, E. O. (2003) Pheidole *in the New World. A dominant, hyperdiverse ant genus.* Cambridge: Harvard University Press.

Wing, M. W. (1968) Taxonomic revision of the Nearctic genus *Acanthomyops* (Hymenoptera: Formicidae). *Memoirs of the Cornell University Agricultural Experiment Station,* 405, 1–173.

INDEX

Page references in **boldface type** indicate main discussions of taxa.

Series Design:	Barbara Jellow
Design Enhancements:	Beth Hansen
Design Development:	Jane Tenenbaum
Indexer:	Jean Mann
Composition:	Jane Rundell
Text:	9/10.5 Minion
Display:	ITC Franklin Gothic Book and Demi
Printer and binder:	Golden Cup Printing Company Limited